Process Development Division 2019

Held at the 2019 AIChE Spring Meeting and
15th Global Congress on Process Safety

New Orleans, Louisiana, USA
31 March - 4 April 2019

ISBN: 978-1-5108-8696-4

Printed from e-media with permission by:

Curran Associates, Inc.
57 Morehouse Lane
Red Hook, NY 12571

Some format issues inherent in the e-media version may also appear in this print version.

Printed by Curran Associates, Inc. (2019)

AIChE
120 Wall Street, FL 23
New York, NY 10005-4020

Phone: (800) 242-4363
Fax: (203) 775-5177

www.aiche.org

Additional copies of this publication are available from:

Curran Associates, Inc.
57 Morehouse Lane
Red Hook, NY 12571 USA
Phone: 845-758-0400
Fax: 845-758-2633
Email: curran@proceedings.com
Web: www.proceedings.com

TABLE OF CONTENTS

549343 Ammonia Manufacturing Box: *Small Ammonia Plant Design Using Micro Processing Technology and Process Intensification Technology*

Ammonia Manufacturing Box

Small Ammonia Plant Design using Micro Processing Technology and Process Intensification Technology

By: Richard B. Strait, Adjunct Professor, Rice University, Department of Chemical and Biomolecular Engineering, Houston, Texas

The small self-contained ammonia manufacturing box (AMB) came out of the need for farmers in remote locations containing modest amounts of natural gas and water to produce enough fertilizer for their local needs.

Presently ammonia plants are getting larger and larger. The future however, is to make small highly automated plants that are portable, use readily available inputs and produce the products where they are needed. To this end the new small scale plant will include all the latest leading edge technologies that apply such as Micro Processing Technology (MPT), Process Intensification (PI) and automated safety systems. The ammonia plant of the future is contained on one rail car, shop fabricated and contains all the latest technology.

Recent developments in micro reactor designs are applied to the ammonia plant reactions; with special attention to: A) steam methane reforming, B) water gas shift and C) ammonia synthesis.

1. Steam methane reforming occurs at high temperatures, is endothermic and an equilibrium reaction. Alternatives are evaluated including: 1) electric reformer, 2) fired heater reformer and 3) auto thermal reformer.
2. Water gas shift occurs at high temperatures with one catalyst for optimum rate and at lower temperatures with a different catalyst to reach a more complete conversion for this endothermic equilibrium reaction. Alternatives are evaluated including: 1) two shift reactors with precise temperature control along the flow path and 2) one reactor with a changing catalyst formulation along the reactor tube length.
3. Ammonia synthesis occurs at high pressures and high temperatures, is exothermic and an equilibrium reaction. Alternatives are evaluated including: 1) precise reactor temperature control following the equilibrium curve and 2) micro channel liquid separation of product ammonia.

Process intensification is applied to CO2 removal, compression, refrigeration and heat integration as well as alternative energy sources.

Safety, operability and maintenance are improved over the traditional stick built plant and these important issues are addressed.

R. B. Strait

October 24, 2018

551619 Building a Continuous Learning and Innovation Network

We live at time when globalization and technology are changing the world in profound ways: economic competitiveness, national security, and personal prosperity are being redefined. At the level of business, hiring companies cannot find people with the necessary skills for today's economy, from truck driving to data mining. Apart from the talent crisis, corporations must reduce new product cycle times and then deliver improved innovations to market, faster. Workers in every industry, meanwhile, find it more and more difficult to climb a career ladder, or even to get onto the bottom rung. For learners of all ages, the rapid evolution in technology and business practices accelerates and intensifies the need for new approaches to career awareness, preparation, and retraining. To help address both human capital development and the need for new knowledge universities and businesses must innovate and think "outside the box" to create new integrated models that lower cost; create customer value; provide a hands-on educational experience and build a foundation for continuous learning. Building a shared / local model that enables transition to the future state is an opportunity for professional societies. This presentation will cover work to date and how we can build globally shared practices – (1) career discovery, (2) integrated learning and (3) market to discovery to enable a new learning and innovation network

550594 Tools for Open Innovation

In today's global and fast-paced environment, the only way companies survive is to innovate. New products and services are required to keep pace with competitors, yet many executives feel that the ideas from within their organizations are stale and incremental.

Disruptive innovation, on the other hand, introduces new ways of solving old problems that engage and empower the customer. The most successful disruptive innovations are built on open innovation platforms, where we involve the customer in the design and development of the new product or service. Of course, there are challenges to managing these relationships, but when successfully implemented, co-creation leads to the highest levels of customer satisfaction.

In this presentation, we will discuss and practice some key tools for open and disruptive innovation. You will learn how to involve a customer in your product design without off-putting jargon or industry buzzwords. In particular, we will discuss the elements of *design thinking* to help you address customer needs and deliver innovations with more accurate product-to-market fit and speed-to-market.

Don't get left behind in a tidal wave of disruption. Learn how to apply open innovation tools to accelerate your success today.

550062 Innovating and Scaling up in Biomass Based Biorefineries – Lessons Learned

The first wave of scaling up biorefineries has encountered many failures and luck luster performance. Such an experience has cast a negative image on the whole industry. Lessons learned about the scaling up of a first of a kind Biorefinery processes will be presented and successful recipes will be discussed.

The open innovation practices of some industries will be compared with the proprietary innovation of the bio refinery industries and contrasts and benefits of the two will be compared.

SAFEHR®: Increasing Safety and Energy Recovery in Sulfuric Acid Plants

NELSON CLARK, BRUNO FERRARO, EDUARDO ALMEIDA, VITOR STURM

Clark Solutions

São Paulo, SP, Brazil

Sulfuric acid is one of the most produced chemicals in the world. Production of sulfuric acid is generally associated with energy production as Sulphur burning and SO_2 to SO_3 conversion are highly exothermic processes taken at high temperatures. These oxidation stages are the main sources of heat in sulfuric acid plants, which is used to produce high quality steam – making energy recovery a fundamental approach for sulfuric acid plants economic viability.

SO_3 absorption into sulfuric acid is another exothermic step which can contribute to overall energy recovery. While there are some successful technologies dedicated to recovering part of absorption energy present in the acid stream, there are still barriers that hold some of these technologies from being standardized solutions. Corrosion, potential shut-down and risks of explosion are some concerns.

During acid-water heat exchanger leaks, a quick and exothermic dilution reaction takes place exponentially increasing corrosion rates and producing hydrogen. In the recent past there were several reported sulfuric acid plant incidents and explosions due to corrosion and hydrogen generation. Some of these failures were catastrophic.

This dangerous scenario can be avoided by means of engineered solutions revolving novel fluid applications. Clark Solutions SAFEHR®, patented Safe Heat Recovery Technology for Sulfuric Acid Plants, is an innovative process that uses inert, immiscible fluids to cool sulfuric acid, virtually eliminating the risk of acid-water contact and its hazards.

SAFEHR creates a new scenario where corrosion and explosion risks associated with boiler or heat exchanger corrosion failures are virtually eliminated while allows the increase of roughly 40-50% in heat recovery and reduction of cooling water requirements in acid plants.

1. Introduction

Energy recovery is fundamental for economic viability in various industry segments. In sulfuric acid plants, Furnace and Catalytic Bed conversion stages are the main sources of heat, which is used to produce high-quality steam using reboiler, superheater and economizer systems. Other substantial amount of heat is generated during SO_3 absorption steps, but most plants do not recover this energy in profitable ways.

The SO_3 absorption is taken inside an absorption tower in which the SO_3 containing gas is under contact with recirculating sulfuric acid, usually fed at 98.5 wt% and 80°C. Due to the highly exothermic reactive absorption, the gas/liquid ratio is designed such as that acid outlet does not exceed 99.5% and 120 °C, preventing excessive mist formation. Prior to tower recirculation, acid concentration is adjusted by dilution, and temperature typically adjusted by cooling water, thus, dissipating a substantial amount of energy.

As sulfuric acid plants generate a large amount of heat, engineers have studied ways to increase heat recovery in the past decades. Whilst heat recovery has increased over time, associated safety has not followed this progress, and increased heat recovery has also added additional operation risk.

Absorption tower heat recovery relies on absorbing SO_3 into highly concentrated acid, 99.0%, at temperatures above 180°C, to produce hot acid stream that could be directly cooled by a boiler, generating saturated 8 to 10 bar steam instead of rejecting heat to the cooling tower. However, a failure of this boiler can be catastrophic; the acid plant and consequently the entire complex is required to shut down, making failure costs escalate. In addition, water-acid reaction at high temperatures is very exothermic and fast, and as reaction takes place, acid dilutes and heats up and, particularly at the leak surroundings, gets more corrosive. The accelerated corrosion process also produces hydrogen gas, introducing an explosion potential to this failure.

This dangerous scenario can be avoided by using technology and engineered solutions revolving an inert, non-corrosive, immiscible, non-toxic, non-flammable and high boiling point intermediate cooling fluid between the hot acid and the steam generation. Thus, making the SO_3 absorption heat recovery process safer. Clark Solutions SAFEHR heat recovery technology addresses these issues, making heat recovery safer than conventional operation where process fluid and water can cause explosions.

Most of sulfuric acid plants, particularly sulphur burning units, typically recover energy from a series of exothermic reactions at the gas side.

As a reference acid plant capacity of 1000 MTPD, the highly exothermic sulphur burning reaction produces almost 127,5 GJ/h of heat for 11% volume of SO_2 at the combustion product.

$$\left.\begin{array}{r} S_{(\ell)} \xrightarrow{445\,^\circ C} S_{(g)} \\ S_{(g)} + O_{2(g)\ in\ dry\ air} \xleftrightarrow{1100^\circ C} SO_{2(g)} \end{array}\right\} \Delta H^\circ \approx -300\,\frac{MJ}{kmol} S \tag{1}$$

The SO_2 to SO_3 conversion, on its turn, also releases another fair amount of heat. As temperature increases with conversion, the system approaches equilibrium, hindering further reaction. Heat needs to be removed to give room for more conversion. This heat amounts to nearly 42,5 GJ/h.

$$\left. SO_{2(g)} + {}^1\!/_2\, O_{2(g)_{remaining}} \xleftrightarrow{420^\circ C} SO_{3(g)} \right\} \Delta H^\circ \approx -100\,\frac{MJ}{kmol} SO_3 \tag{2}$$

The heat generated on the Sulphur and Sulphur Dioxide oxidation is usually recovered as high-pressure steam (HPS). Several economizer/boiler/superheater arrangements are used to recover this energy, with small variations in extent of energy recovery and temperatures in the system.

As an average, double absorption acid plants will generate something from 1.1 to 1.3 massic tons of HPS per ton of acid produced. Single absorption plants, despite the lower SO_2 conversion efficiency, will produce more HPS, from 1.4 to 1.6 ratio. And plants with acid heat recovery can reach as high as 1.8 to 2.0 ratio.

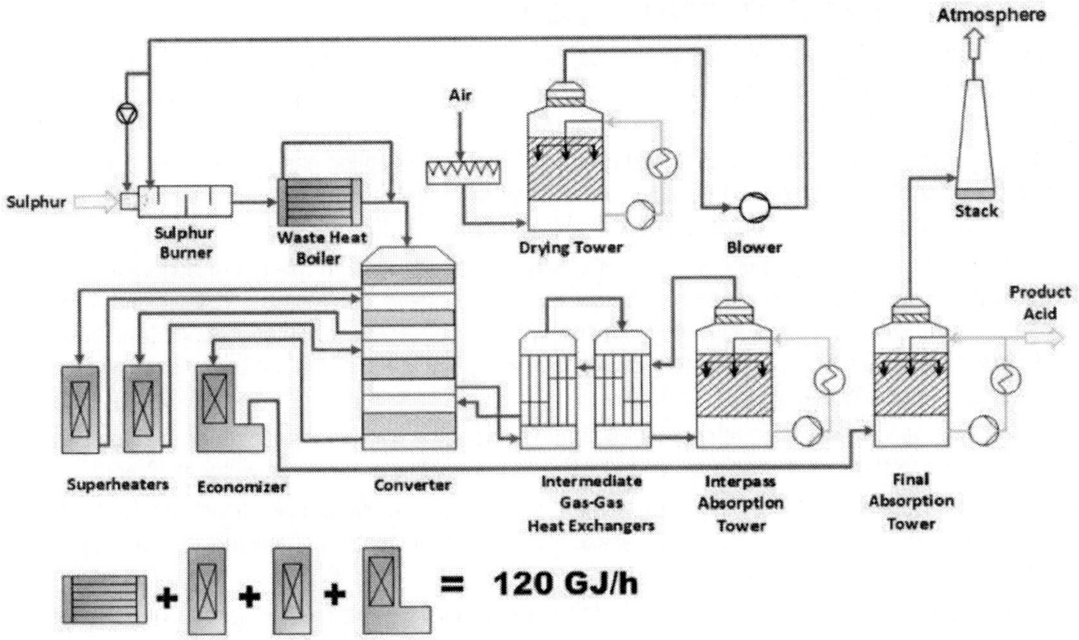

Fig. 1: Conventional 1000 MTPD Acid Plant and Gas Energy Recovery

Despite generating roughly 170 GJ/h of heat at the gas, not all energy is recovered due to efficiency and thermodynamic limitations, as well as process restraints. For instance, a higher dewpoint achieved at the Drying Tower (higher water content in gas) brings a higher acid condensation temperature on heat exchanger surfaces. For this reason, gases from conversion are cooled to a minimum between 150°C to 200°C, and sometimes as high as 300°C, to avoid acid condensing on economizers – and this specific non-cooled heat is transferred to absorption towers irrigated acid and subsequently to cooling water.

Besides heat generated on gas steps, another important heat generation takes place during SO_3 absorption inside acid towers, especially in the Interpass Absorption Tower (IPAT).

$$SO_{3_{in\ gas}} + H_2O_{(\ell)_{in\ 98,5\%wt.\ H_2SO_4}} \rightarrow H_2SO_{4\,(\ell)}\Big\}\ \Delta H^\circ \approx -130\frac{MJ}{kmol}H_2SO_4 \qquad (3)$$

On our hypothetical 1000 MTPD acid plant, all absorption energy would amount to approximately 63 GJ/h of heat, of which up to 85% (52 GJ/h) can be solely generated in the IPAT, depending on the conversion arrangement. The total heat generated in a 1000 MTPD plant is nearly 233 GJ/h, or 65 MW.

Of the total heat released in the process, the conventional technology, which leads the choice and model of construction on most modern sulfuric acid plants, only consistently recovers energy from burning and conversion, discarding all absorption energy. Nearly 25% of the total heat produced in the plant is wasted alongside cooling water.

In the mid 80's Monsanto Enviro-Chem Systems proposed a new approach to increase heat recovery in acid plants, the HRS – Heat Recovery Systems (U.S. Patent No. 4,576,813, 1986).

The HRS recovered heat from SO_3 absorption reaction inside the IPAT. Instead of absorbing SO_3 at standard temperatures (70°C to 80°C), Enviro-Chem proposed the absorption at higher temperatures (180°C to 200°C) followed by cooling hot acid on low pressure boilers (10 bar and bellow) to recover energy that otherwise would be discarded on water cooling towers.

The HRS was an important, if not the most important, improvement in acid production since the application of the double absorption patent by Bayer in 1960 (U.S. Patent No. 3,259,459, 1960).

2. Recovering Heat from Hot Acid

The absorption of SO_3 in the contact process can be done in a single stage (as in the originally conceived single absorption plants) or in two steps, with one intermediate absorption step after bed #2 or bed #3 and a final step prior to stack release.

In the single absorption configuration almost (97+%) of all SO_3 absorption energy can be recovered while in the double absorption acid plant approximately 75-90% of the absorption energy can be recovered, depending on the conversion arrangement of the plant (2/2, 3/1 or 3/2).

On the conventional arrangement, where absorption acid is admitted at the top of absorption towers at temperatures below 85°C and the bottom acid temperatures are controlled to not exceed 115°C, the temperatures do not allow any quality energy recovery. In some places, hot water obtained from acid cooling acid is used for city district heating or for electrolyte heating, but the temperature does not allow the energy to be recovered as any useful form of steam.

When one increases absorption temperatures, energy can be recovered at a higher grade, such as low or medium pressure steam (IPS). Acid at 200 to 225°C can easily boil pressurized water at 10 bar.

Besides increased heat recovery, high temperature acid absorption offers a significant additional benefit: it reduces cooling water consumption and associated treatment costs, an advantage that is especially beneficial for plants with limited access to water.

But these benefits come associated with a definitively non-negligible risk: since heat recovery takes place on an acid/water boiler, a failure or leak in the system can, and a few times did, proved to be disastrous.

It is believed that the associated risk with this steam and hot acid system has historically been an important deterrent to the widespread and standardization of heat recovery from hot acid in sulfuric acid plants.

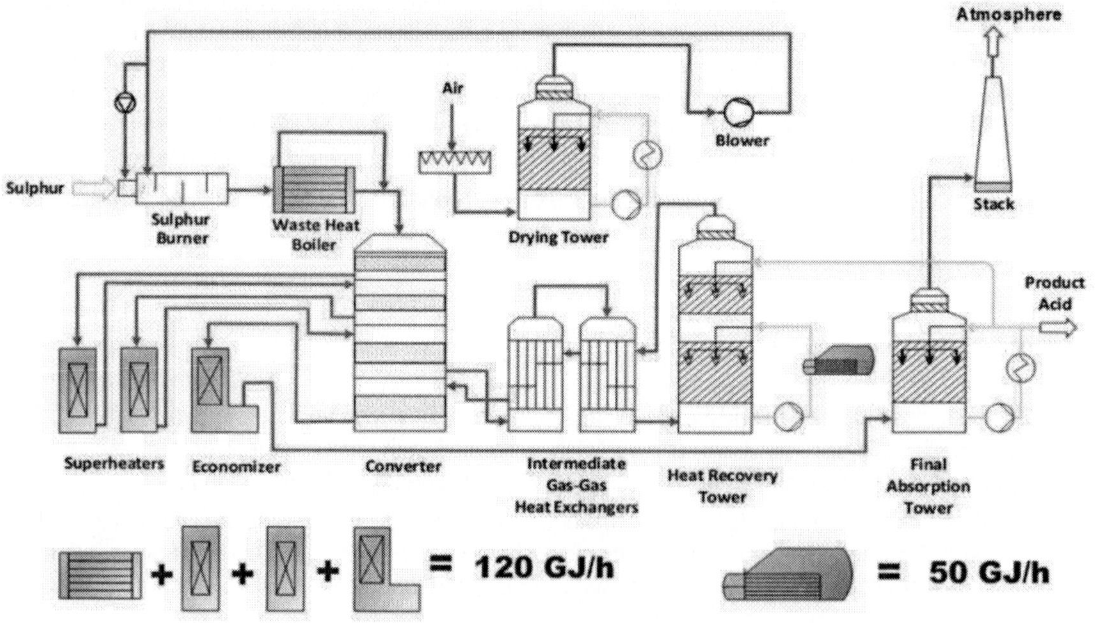

Fig. 2: 1000 MTPD Acid Plant with Gas and Acid Energy Recovery

Clark Solutions SAFEHR technology actively address this safety issue and minimizes, if not eliminates, the risks associated to the steam/acid system leakage.

2.1. Safety and Hydrogen Explosion

The sulfuric acid industry has been recently emphasizing hydrogen explosion hazards. Hydrogen gas is formed as a byproduct of metal corrosion by sulfuric acid and can be accumulated inside plant equipment. Weak sulfuric acid is very aggressive and when present, hydrogen is generated at high rates, enough to reach its explosive limits since energy ignition energy for this gas is very low.

$$
\begin{aligned}
&Fe^{0}{}_{(s)} \xrightarrow{yields} Fe^{2+}{}_{(aq.)} + 2e^{-} \\
&2H^{+}{}_{(aq.)} + 2e^{-} \xrightarrow{yields} H_{2(g)} \uparrow
\end{aligned}
\tag{4}
$$

Weak acid condensation due to moisture presence and dilution of strong acid due to cooling water leaks, among others, can bring dangerous scenarios. In case of water leaks, the risks can be worse, since the dilution itself is exothermic, and the heat generated speeds the corrosion process.

Several hydrogen explosions have been reported by the industry in the recent past.

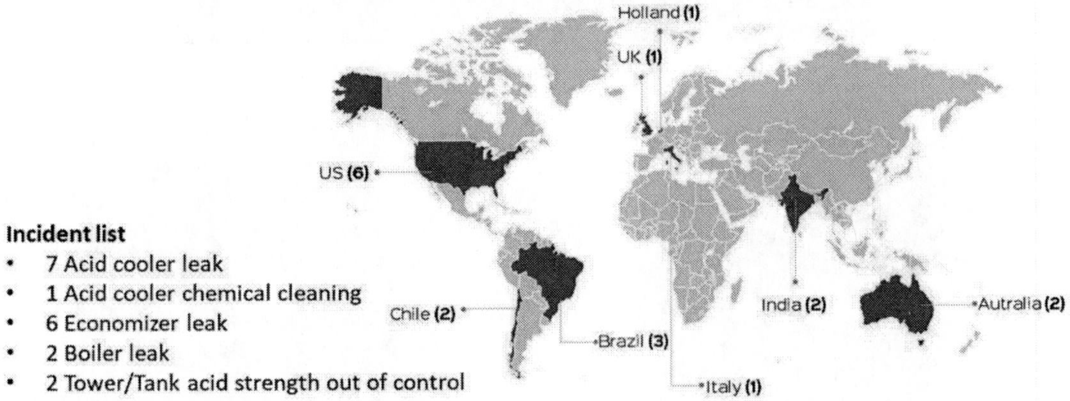

Fig. 3: Hydrogen Explosions Incidents – Incomplete data as of 2015

On a conventional acid cooling system in a plant that does not recover SO_3 absorption energy, acid coolers flow is undertaken in a condition that acid pressure is always higher than cooling water pressure. In the event of a failure, acid will leak into the water side allowing quick identification by simply measuring water pH downstream from coolers and avoiding dilute acid inside plant equipment at the expense of minor damages to the cooling water system, which is typically less expensive and easier to handle.

On the other hand, on a conventional heat recovery system acid is always at a lower pressure than boiling water, which is pressurized at least from 4 to 10 bar, depending on steam demand.

The large pressure difference between acid and water is a stronger driving force than the pressure difference in a conventional system. Therefore, a leak either initiated by corrosion, by bad material or workmanship quality or even a stress issue will occur at substantially higher flow rates than that in a conventional system. Similarly, acid dilution and heat release rates will become higher. Ultimately leading to critical failures.

Additionally, this leak region receives hot acid at 180+°C being mixed by hot water to produce a dilute acid at even higher temperatures. Local conditions are tremendously aggressive.

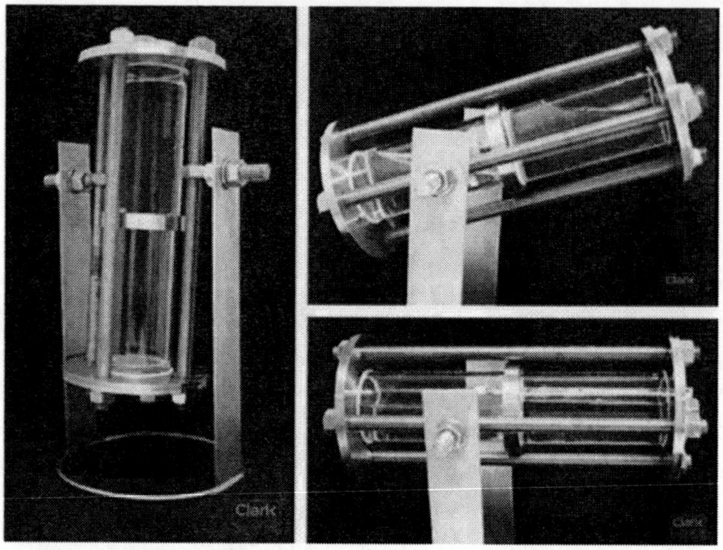

Fig. 4: Scheme of corrosion test apparatus.

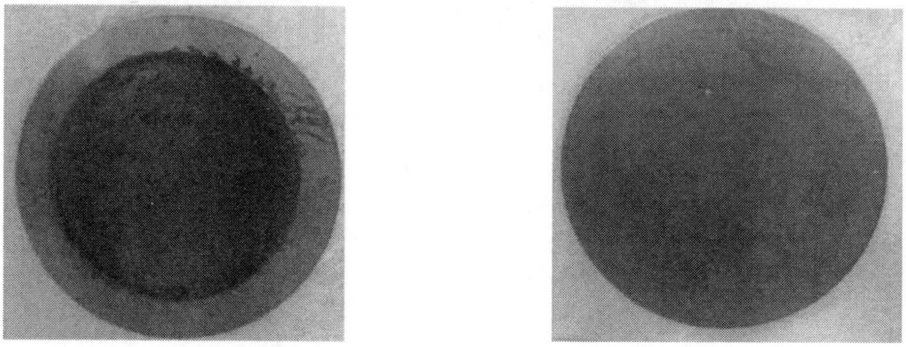

Fig. 5: Comparison of 316L corrosion for acid/water mixture (left) and acid/fluid mixture (right).

Studies of corrosion progression developed by Clark Solutions show that the corrosion rate of a 1 mm hole in an acid-water system at ambient temperature and pressure can be as high as 150 mm per year, with the hole diameter growing as much as 0,1% per hour in these mild conditions.

Table 1
Sulfuric Acid Corrosion Progression on Leaking Hole

Fluid Pair	Conditions	Results
Water 98% Sulfuric Acid	12 hours duration Start at ambient temperature 316L stainless steel apparatus	**Corrosion rate: 145 mm/year Mass lost: 0,1% per hour**
CS270 Inert Fluid 98% Sulfuric Acid	72 hours duration Start at ambient temperature 316L stainless steel apparatus	**Corrosion rate: nil** **Mass lost: nil**

When in a heat recovery system, a water leak into the acid can dilute the strong acid in the system. The corrosion effects will occur inside the acid plant equipment, something that does not normally occur on a leak in a conventional system, unless in some specific maloperation or improper design situations. The corrosion in turn will release hydrogen inside plant equipment. Hydrogen will accumulate in high spots. This may further increase operation risk.

Several actions and measures have been taken to minimize the risk of a disastrous failure in a heat recovery system due to water into acid leak.

From modern, sophisticated and redundant concentration, corrosion and temperature digital monitoring and control to segregated heat recovery towers such as Venturi have been employed to reduce emergency shut-down probability.

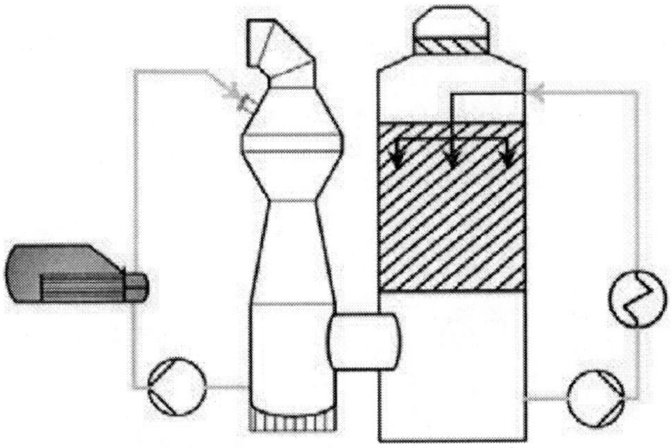

Fig. 6: System with segregated Venturi for energy recovery

The Venturi irrigation can be shut-down allowing the system to operate as a standard absorption tower.

Advances in instrumentation has also substantially reduced risks with monitoring and safety measures. However, none of these safety schemes eliminate the root cause of the failure: acid-water contact.

2.2. Indirect Cooling of Hot Acid

In 2014 Clark Solutions has been contacted by a large Brazilian fertilizer company interested in increasing heat recovery by means of a boiler feed water (BFW) pre-heat system for its large 3200 MTPD sulfuric acid plant. The project should be done in such a way to avoid putting BFW quality at risk and avoiding plant shut-down in the event of a BFW heater system failure.

This customer had experience with the existing boiler feedwater systems installed in Brazil and was seriously concerned about risks.

All the existing installed BFW pre-heater systems relied into a shell and tube heat exchanger that would cool hot acid by direct contact with BFW. In the case of a failure not only BFW could get contaminated but the plant would usually have to immediately shut down.

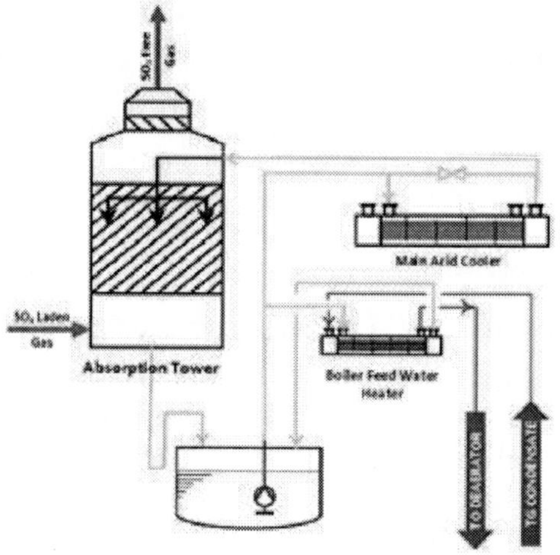

Fig. 7: Conventional BFW Heater with Shell and Tubes

Plate coolers were selected for this project. Their use brings some benefits when compared with shell and tube exchangers: (1) instead of a 10 to 15°C temperature approach achieved in shell and tube exchangers, plate exchangers achieve 3 to 5°C approach, improving recovery and (2) higher grade material plate exchangers reduce overall cost due to smaller areas offered by the higher heat transfer coefficients, keeping in mind that anodically protected shell and tubes passive operation is limited to surface temperature.

Clark Solutions approach was to use a trim system with a two plate coolers arrangement and an intermediate water circuit. Acid would be cooled by a closed-circuit high-quality low fouling water and this heated water would then transfer acquired acid heat to the BFW, increasing its temperature from 50 to roughly 100°C.

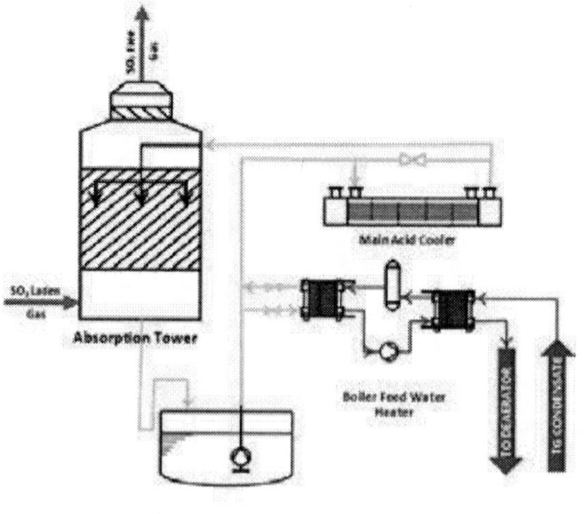

Fig. 8: BFW Heater with Plates and Closed Circuit

The system was self-contained, fully skidded and instrumented, so in the event of a leak it could be taken out of service, diverting both acid and BFW streams from the system without requiring plant shut-down.

The successful solution brought up the idea of cooling not only a small acid fraction for BFW pre-heating, but also using all absorption acid flow at even higher temperatures in a similar arrangement to produce steam.

But how to avoid the same risks experienced by the water filled boilers and heat exchangers in similar services?

2.3. Inert Fluid

In the early 2000's Clark Solutions has done extensive testing and evaluation of direct cooling of sulfuric acid by mixing with inert fluids. On this research it was found that a few families of fluorinated hydrocarbons had several adequate properties:

- Stability
- Inert to acid and water
- Non-flammable
- Non-Toxic
- Low Vapor Pressures
- Safe to handle
- Substantial density difference between fluid and water/acid
- Good heat transfer properties

The direct contact approach failed; the amount of fluid required was substantially high, fluid final temperature would be much lower than the acid temperatures – impeding any heat recovery – and the potential loss of fluid due to acid carryover on the acid stream could be costly.

While the direct cooling did not prove commercially attractive, using the inert fluid would be an excellent approach for a trim cooling system: a leak on the acid heat exchanger would not increase due to the heating/dilution effect, no contamination of acid or water inventories, and a controlled leak would allow operators to plan a maintenance shut-down, which otherwise would happen in an immediate emergency state as the uncontrolled fast increasing leak growth would be catastrophic.

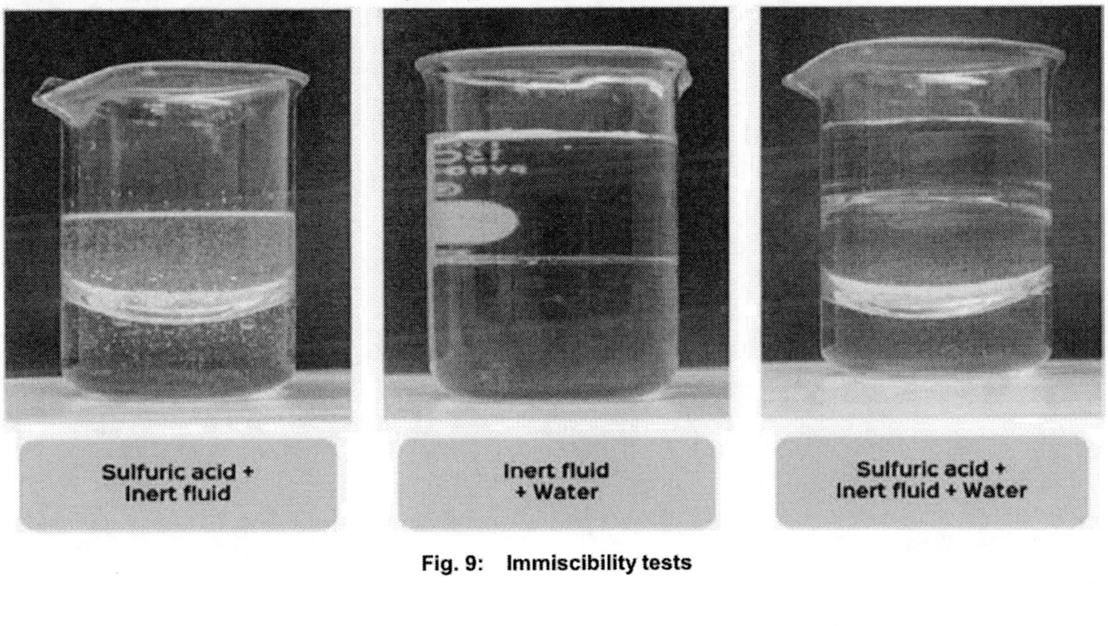

Fig. 9: Immiscibility tests

3. A SAFEHR® Approach

SAFEHR is the concept of cooling hot process fluid with an inert fluid indirect trim cooler loop.

Its fundamental purpose is to increase operation safety, minimizing damages due to leaks and reduce hydrogen generation due to corrosion. An additional benefit is the possibility of increasing heat recovery and HPS/IPS generation.

The SAFEHR system is basically comprised of two heat exchangers using a properly selected inert fluid in an enclosed system. Process fluid is cooled on the first exchanger by heating the inert fluid which on its turn is circulated in the enclosed system and cooled by heating BFW or another fluid.

SAFEHR can be used on a conventional cooling system as a safety improvement, in replacement for conventional coolers, or it can be used attached to a heat recovery tower, in replacement to the boiler/heat exchanger.

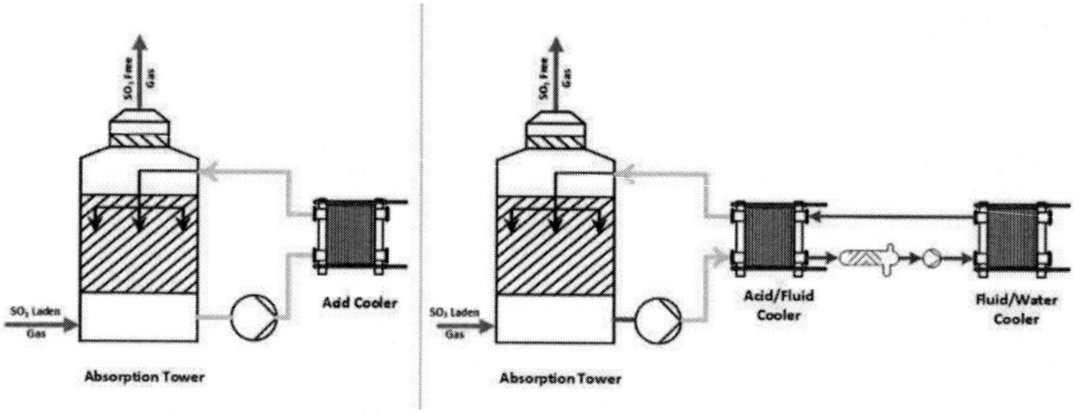

Fig. 10: Conventional cooling (left) and SAFEHR® cooling (right)

Or even as a complete solution, with heat recovery tower and cooling supplied as part of a system.

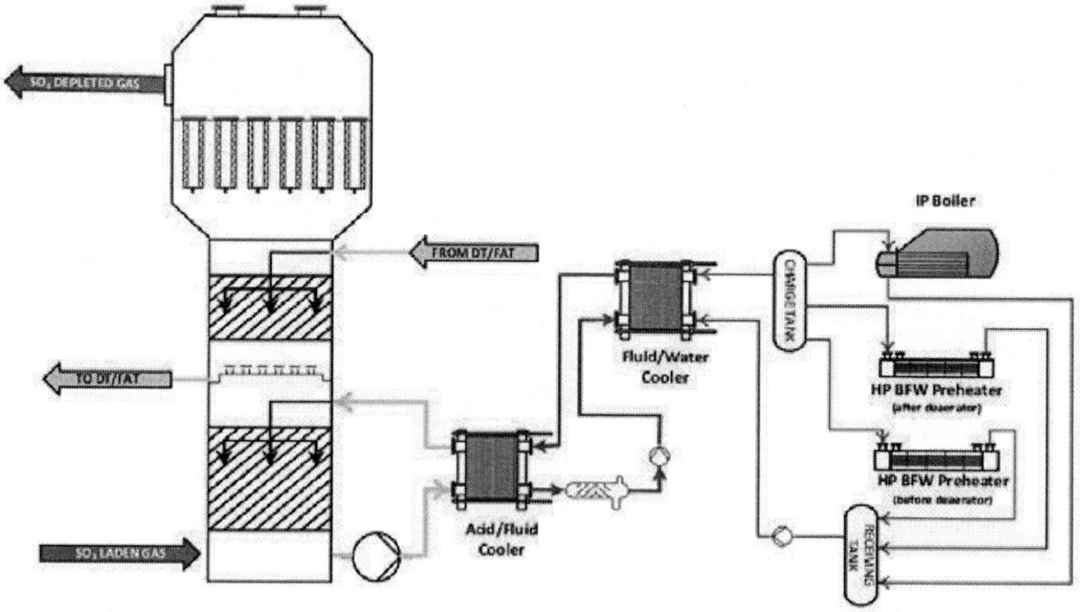

Fig. 11: SAFEHR cooling and energy recovery. Tower bottom section operates at 180 to 220°C and manifold allows SAFEHR to redistribute energy for either low- or high-pressure steam generation or utilities

Another benefit from SAFEHR, especially in sulfuric acid plants, is that the SO_3-laden gas from conversion is no longer required to be cooled as low as 150°C on economizers to save energy, as its energy will already be transferred to SAFEHR steam generation. This way, smaller economizers can be employed and lower risk of acid condensation on gas equipment and pipes is achieved, reducing another risk of corrosion and failure.

Fig. 12: 3D of SAFEHR Heat Recovery System

The SAFEHR system is also comprised by pumps to circulate inert fluid, an expansion vessel to absorb fluid expansion and a liquid-liquid coalescer where any leak can be detected and segregated from the system.

The liquid-liquid coalescer is a cost-efficient separation approach. Immiscible suspended droplets can be grown and efficiently separated into two films in properly designed coalescers.

SAFEHR is assembled to flow process fluid into the trim system when leaks occur. This causes no damage as process fluid and inert fluid are immiscible. For instance, a dispersed leak of sulfuric acid into inert fluid, when properly coalesced, will form a descending film inside coalesce vessel and fall towards its bottom, thus detected, and removed from the system via a trap boot.

In the event of a water leak into the system, water will float and increase system pressure, allowing easy detection by pressure gauge.

In both events, water or acid leaks, the latter more probable, the plant does not require a shut-down. The leaks will not increase as in the acid/water systems as synergic effects of heat and dilution do not take place.

Instrumentation, on opposition to the conventional heat recovery systems, is simple, as a leak will not become a catastrophe trigger. Simple pH, temperature and level controls are enough for a robust and safe control.

The trim intermediate system can be employed in replacement to any heat exchanger system where cool and hot media eventual mixture can cause severe damage. The trim media has a drawback to temperature approach, and heat exchangers must be selected accordingly to maintain necessary temperature profiles.

SAFEHR system heat exchangers can be selected and designed accordingly to process conditions and process fluid. The first high temperature commercial unit utilized fully welded block exchangers for a compact design capable of handling higher pressures. The Alfa Laval Compabloc, especially the latest 310S model developed in France for the SAFEHR application, showed the best cost-effective solution with higher temperature approach and smaller closed-circuit, reducing costs in piping and in inert fluid, and overall system cost. At the inert fluid/water side, the exchanger can also be a kettle boiler.

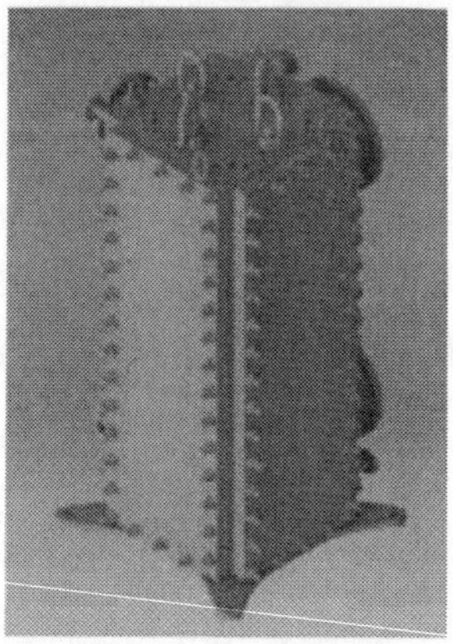

Fig. 13: 310S Welded Block (Compabloc) Heat Exchanger for Energy Recovery and Steam Generation

As a closed circuit, the inert fluid system should be designed to maintain adequate pipe velocity and heat exchanger pressure drop. A system retrieving as high as 5 MW of thermal energy can easily fit into a container sized skid.

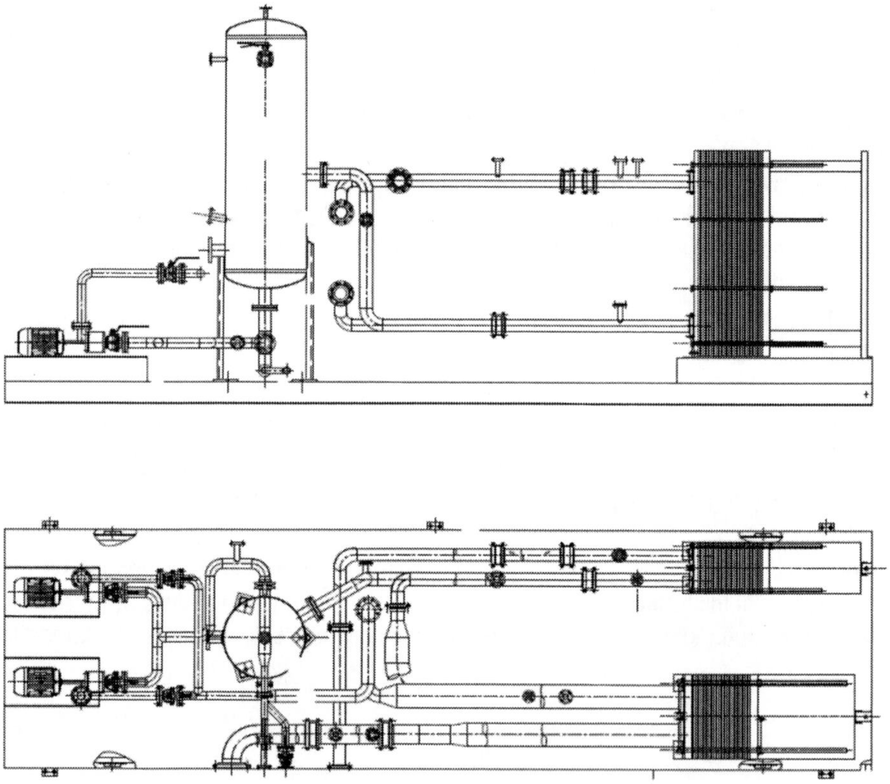

Fig. 14: Scheme of closed-circuit skid for BFW pre-heating with plate heat exchangers

4. Materials of Construction

Materials of construction for use in strong acid services have been extensively studied.

Strong acid at normal absorption temperatures can be handled with a myriad of materials that proved efficient for such conditions. High Silicon Austenitic Stainless Steels like CSX™ and similar (CSX™ is the trademark for Clark Solutions high silicon stainless steel), 310 Stainless Steel, Alloy 20, Cast Iron and Anodically Protected 304/316 Stainless Steel have all proven resistant to the common some range of acid conditions.

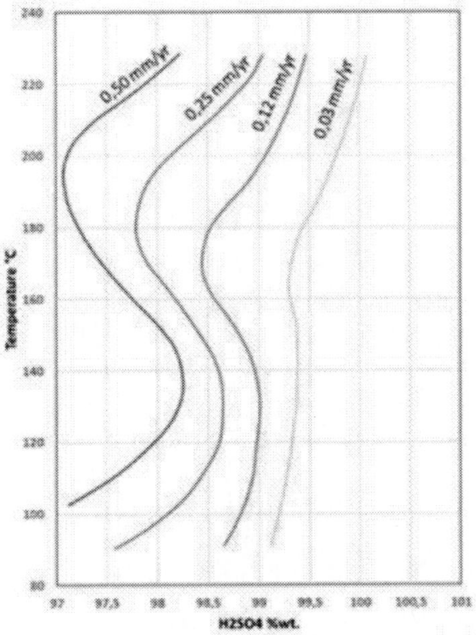

Fig. 15: 304L Stainless Steel Isocorrosion Lines

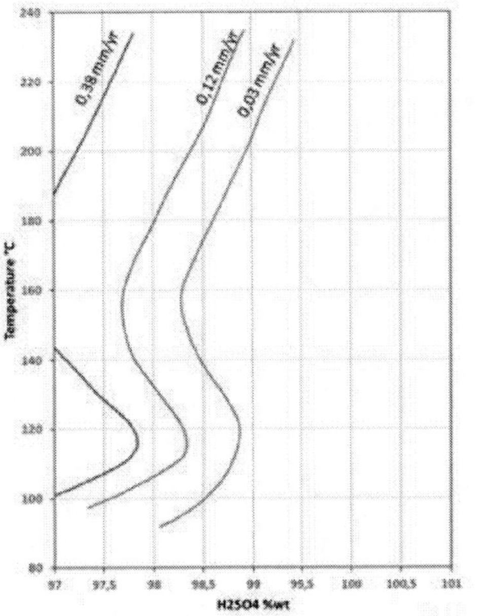

Fig. 16: 310S Stainless Steel Isocorrosion Lines

When acid is circulated at higher temperatures and concentrations (200+°C / 99.0+%) the same materials that perform well at lower temperatures do not offer the same corrosion resistance.

When subjected to higher concentration and temperatures, neither the high Silicon austenitic steels nor 316L are any longer a good option.

McAlister and Ziebold (U.S. Patent No. 4,670,242, 1987) found that among the best cost benefit options for heat recovery applications are 304L and 310, even though many other materials also present very good performance in the operating conditions.

While proper material selection is very important to minimize failure risks, it still does not eliminate risks. Corrosion is a matter of time and operational control – and leaks eventually happen.

SAFEHR systems can be constructed in the materials suitable to the operating conditions; for instance, if the operating concentration and temperatures of the acid are in the 98.0 to 98.5% and 100°C to 115°C range, Hastelloy D-205 (plate) or CSX (S&T) or anodically protected 316L (S&T) can be used in the main acid/fluid cooler.

If the acid conditions are those of a heat recovery system, for instance, in the 99.0 to 99.5% and 200°C to 220°C, the main acid/fluid cooler can be either 310S Stainless Steel (S&T and Welded block) or Alloy 33 (S&T and Plate).

The water/fluid cooler material can be carbon steel, 304L or 316L. The choice will depend on the system arrangement, if the exchanger is boiling water or not, or customers budget as, on normal conditions, this exchanger will never see hot concentrated acid.

In addition to the materials of construction for the tower, piping and exchangers, Clark Solutions has done research on materials for pumps and valves, particularly in the high temperature/high concentration cases. Partnerships with major pump, valve and accessories have been made to guarantee quality long-life materials to the installations.

5. SAFEHR Performance

Even though the technology was conceived to increase safety, we found that SAFEHR will add some extra benefits to operation when compared to conventional heat recovery systems.

If the system is designed to high pressure service, it can be used to pre-heat BFW to temperatures up to 180°C to 210°C, this will increase the HPS generation by 15%, making HPS the major product of the SAFEHR System.

Given the enormous amount of energy present on the acid, LPS/IPS will still be produced, but at a lower rate than in a conventional heat recovery system.

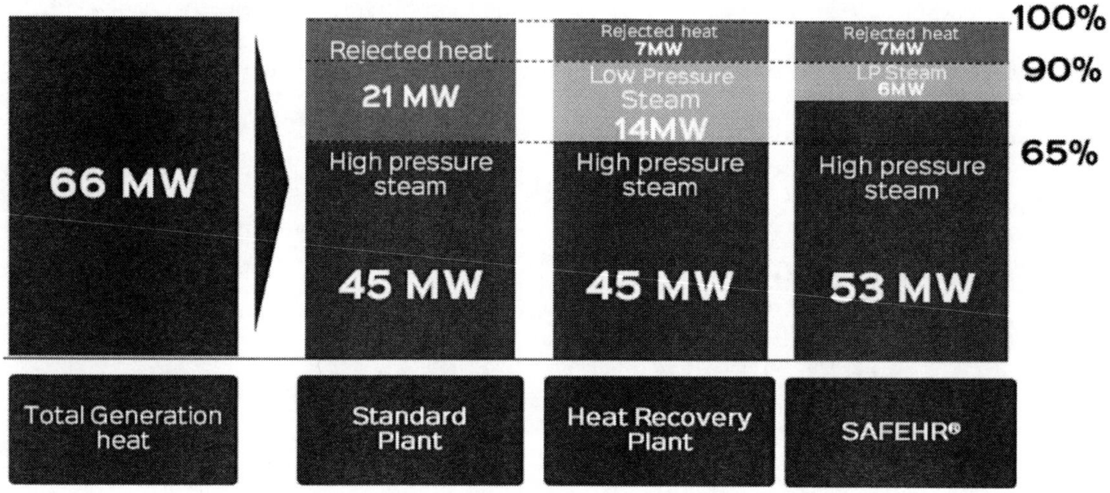

Fig. 17: Energy breakdown in different configurations of a 1000 MTPD plant

The use of heat recovery system comes with secondary cost benefits, such as reduction of cooling water capacity. This implies in reduction of water treatment costs and water make-up. The reduction of circulating cooling water in the plant can be as high as 0.3 m³/h (per ton of produced acid per day) if the high-pressure steam is used to generate energy in a turbo generator with 25% efficiency – live steam use on site operations can increase economy in circulating water as there is reduced steam condensing requirements.

As an operational benefit, since the high-pressure water will get hotter, the tube wall temperature in interpass/single absorption tower economizer will be higher, preventing condensation on this critical equipment, increasing plant uptime reliability in one of the highest maintenance pieces of equipment.

SAFEHR can also be operated in a reverse cycle. That is: hot water heats the intermediate fluid that in turn heats acid. Reverse cycle reduces required time for start-ups.

6. The First Plant

The first SAFEHR high recovery unit is a modular 150 MTPD single absorption acid plant with a hydrogen peroxide tail gas scrubber. It will produce strong acid and IP steam for leaching operations at an aluminium-potassium mining.

Fig. 18: 3D Layout of modular small capacity plant (150 MTPD)

7. Conclusion

SAFEHR is a new approach that adds a substantial safety feature to existing or new sulfuric acid plants. This system can be applied to other industries where process fluid and water unexpected mixture can potentially cause equipment damage and explosions, such as phosphoric acid or hydrogen chloride heat exchangers. For the discussed case, energy recovery from SO_3 absorption heat is achieved at a safer operation with reduced hydrogen generation risks, and parallel operational benefits as it allows for lower acid condensation rates on economizers and other gas heat exchangers. All of this achieved with an increased production of high-quality high-pressure steam.

This shows that novel safer technologies can also increase operational income.

References

Clark, N.P. (2017). Sulphur, 373, November-December 2017, pages 53-57.

Clark, N.P.; Ferraro, B.B.; Sturm, V.A. (2018). 34th Sulphur + Sulphuric Acid International Conference & Exhibition. Gothenburg, Sweden.

McAlister, D. R.; Ziebold, S. A. (1986). *United States Patent No. US4,576,813*. St. Louis, MO: U.S. Patent and Trademark Office.

McAlister, D. R.; Ziebold, S. A. (1987). *United States Patent No. US4,670,242*. St. Louis, MO: U.S. Patent and Trademark Office.

Wilhelm, M. (1960). *United States Patent No. US3,259,459*. Leverkusen, Germany: U.S. Patent and Trademark Office.

Acknowledgement

We would like to thank Luís Lacombe and Ricardo Dequech from Kalium Mining for embracing this technology.

550286 Controllability Studies for CO_2 Post-Combustion Capture Processes By Dynamic Simulation

CO_2 is the major greenhouse gas that is widely condemned as the root cause of the global climate change. The emission of CO_2 from fossil fuel combustion and industry processes contribute 78% of total greenhouse gas emission due to the globally economic and population growth. Therefore, the continuous technology advancement for CO_2 capture are crucial to reduce global greenhouse gas emissions as well as global climate changes. Amine-based post combustion CO_2 capture process has a long application history and has been recognized as a mature technology for coal-based power plant. Many studies have been conducted to improve this technology, such as choosing better amine solvents, optimize the lean solvent inlet conditions, modification the typical process designs and so on. However, few of the reported studies addressed dynamic process upset issues during the real practice. In applications, the intake flow of waste CO_2 streams will most likely fluctuate a lot due to the peak and off-peak of operation loads, which may lead to significant upsets for the CO_2 capture process, and thus affect the process efficiency and product quality.

In this study, plant-wide steady-state and dynamic models of a CO_2 post-combustion capture process with monoethanolamine (MEA) and Methyl diethanolamine (MDEA) blended amines have been developed to study the operational flexibility and controllability under disturbances. Modeling and simulation results shows the flowrate and temperature of lean loading stream entering the absorber and stripper's reboiler duty are critical factors for CO_2 absorption performance. Therefore, novel control strategies with these factors by using advanced process control (APC) are studied in this work, this could not only maintain the feasible operation of the system, but also improve the energy performance under disturbances. Final results show that the fluctuate of CO_2 removal efficiency is no more than 0.1%, while the inlet flow disturbances have at most 30% changes during this period. Better control strategies could save extra 7% energy from heat exchangers and 18% power from pumps, which significantly improve the stability and economic performance of the studied CO_2 capture process.

550545 A Comparison of the Decompositions of Benzene and Toluene As Tar Representatives Using Non-Thermal Plasma

Non-thermal plasma is being considered as an alternative technology to decompose gasification tars. In this study, a dielectric barrier discharge (DBD) reactor was used to remove the tar representatives (benzene and toluene) in a model product gas (CH_4, CO_2, CO and H_2). The effect of specific input energy (SIE) (2.05-16 kWh/m^3), residence time (0.95-2.82s), and temperature (20-400°C) were studied. The decomposition of benzene and toluene increased with increasing SIE and residence time. At ambient temperatures, the formation of lower hydrocarbons, and solid residue occurred. This would be a significant operational problem, but here we show that the effect can be significantly reduced by operation at higher temperatures. In the presence of plasma at 400 °C, solid formation was substantially reduced. At the same time, selectivity to lower hydrocarbons increased from 31 to 74 % and 36 to 79 % for toluene and benzene, respectively. However, the decomposition of CO_2 and CH_4 decreased with increasing the temperature.

550506 Grams - a Dynamic Intensification and Optimization Platform for Modular Chemical Process Systems

A significant portion of future demands for chemical feedstocks is predicted to be met by unconventional and distributed resources such as associated shale gas, stranded natural gas, coalbed methane, landfill gas, biogas, and fuel gas. Current roadblocks to widespread utilization of these resources often include a lack of pipeline infrastructure, sparsity, seasonal variability and uncertainty in feedstock volumes and compositions, and occasional high levels of contamination. Small-scale, modular chemical process systems (MCPS) can be a promising solution to tap into these unconventional resources. They offer increased flexibility, agility and ability to accommodate varying market demands. However, poor economies-of-scale and high capital intensity prohibit most small-scale MCPS to be economically viable over their large-scale counterparts. We argue that this limitation can be overcome through dynamic process intensification that combines multiple phenomena, such as separation, conversion and intermittent storage, within a single equipment in a periodic manner. One example of such a multifunctional MCPS technology is the sorption-enhanced reaction process (SERP). In SERP systems, *in situ* capture of undesired reaction byproducts by adsorbents favors the desired reactions in catalysts, thereby increasing the overall yield of desired products. To this end, advanced modeling, simulation and optimization of such dynamically intensified and complex systems can provide critical breakthroughs needed for small-scale MCPS towards the economic production of chemicals utilizing unconventional feedstocks.

We have developed GRAMS (Generalized Reaction-Adsorption modeling, optimization and Simulation), which is a first-of-its-kind computational framework for the optimal design and operation of dynamically intensified packed-bed systems. For given feeds and products specifications, GRAMS is used to optimally design the cycle configurations, column design specifications and process operating conditions of a periodically operated MCPS [1-3]. The GRAMS platform combines an in-house dynamic process simulator with a cost estimator and a data-driven constrained grey-box optimizer [4]. The high-fidelity process simulator is based on a first-principles-based model of a generalized reaction-adsorption system to predict the performances of processes incorporating periodic gas adsorption-desorption (e.g., pressure/temperature/vacuum swing adsorption: PSA, TSA, VSA, SMB), reaction (e.g., PFR, SMBR), or a combination of both (e.g., SERP). The model predictions are extensively validated with experimental data for industrially relevant pressure swing adsorption (PSA), steam methane reforming (SMR), methanol synthesis, sorption-enhanced SMR (SE-SMR), and sorption-enhanced water gas shift reaction (SE-WGSR) processes. The framework has been used for the optimal synthesis of three multi-mode, multi-step and periodic SERP systems, namely SE-SMR, SE-WGSR and sorption-enhanced methanol (SE-MeOH). The optimized SE-SMR produces hydrogen from natural gas with 35% higher productivity and more than 10% lower cost in comparison to existing small-scale systems [2]. Furthermore, the novel SE-MeOH process, designed using GRAMS, could lead to more than 7% improvement in methanol yield with only 2% decrease in production capacity [3].

References

[1] A. Arora, S. S. Iyer, and M. M. F. Hasan, "GRAMS: A General Framework Describing Adsorption, Reaction and Sorption-Enhanced Reaction Processes," *Chem. Eng. Sci.*, vol. 192, pp. 335–358, 2018.

[2] A. Arora, I. Bajaj, S. S. Iyer, and M. M. F. Hasan, "Optimal Synthesis of Periodic Sorption Enhanced Reaction Processes with Application to Hydrogen Production," *Comput. Chem. Eng.*, vol. 115, pp. 89–111, 2018.

[3] A. Arora, S. S. Iyer, I. Bajaj, and M. M. F. Hasan, "Optimal Methanol Production via Sorption Enhanced Reaction Process," *Ind. Eng. Chem. Res.*, 2018.

[4] I. Bajaj, S. S. Iyer, and M. M. F. Hasan, "A Trust Region-based Two Phase Algorithm for Constrained Black-box and Grey-box Optimization with Infeasible Initial Point," *Comput. Chem. Eng.*, 2017.

550635 Low-Temperature Fischer – Tropsch Synthesis Multiphase Product Recovery through Reactive Distillation

The Fischer–Tropsch (FT) process relies on a catalyzed chemical reaction to convert synthesis gas into paraffins, olefins, and an appreciable amount of oxygenates. The synthesized FT product slate is remarkable, ranging from C_1 (methane) up to C_{70+}. The distribution of FT products varies based on reaction conditions (i.e., temperature, pressure, space velocity, and H2/CO ratio), reactor type, and catalyst formulation. We have developed an equation-oriented framework for optimal synthesis of integrated reactive distillation (RD) systems for FT processes in GAMS in which the phase equilibrium is given by a cubic equation of state, the reaction rate expressions are expressed in terms of fugacities, and the product selectivity of catalysts are based on the Anderson–Schulz–Flory distribution and experimental data [1].

In this work, we are expanding the model to consider a more comprehensive kinetic expression with a wider product selectivity to explore the potential of producing a predominantly wax product on a low-temperature FT process, with an expanded model to optimize tray numbers, feed tray location and catalyst loading on trays. We are also implementing the FT-RD optimization model within the PyOMO optimization framework.

The results show that exothermic heat of reaction is sufficient to provide enough vapor load which renders a reboiler unnecessary; and side-streams of high quality gasoline and diesel can be obtained. Thus demonstrating that novel reactive distillation processes are paving the way for a more sustainable chemical process industry that is profitable, safer and less polluting.

[1] Y. Zhang, C.M. Masuku, L.T. Biegler, Equation-oriented framework for optimal synthesis of integrated reactive distillation systems for the Fischer–Tropsch processes, Energy & Fuels 32 (2018) 7199–7209.

551084 Modeling and Simulation of 1,3-Butadiene Production Process from Lignin

1,3-butadiene is utilized in a wide range of industries such as healthcare, automotive, building and construction and consumer durables [1]. Almost all 1,3- butadiene is produced from steam cracking of heavy crude feedstock. As the cracker feedstock around the globe is trending towards the lighter feedstocks because of the less expansive ethane produced from shale gas, the sustained production of 1,3-butadiene in olefin plants is facing big challenges. As per the current market trends in North America and Europe, 1,3-butadiene production is expected to continue to decrease for next several years [2].

To reduce this gap between demand and supply of 1,3-butadiene, an alternative feedstock for butadiene production is required. In this research, we study the feasibility of using Lignin as an alternative source for 1,3-butadiene production via simulations. First, Lignin is gasified and converted to syngas. Then, the conversion of syngas to light olefins is studied in three different process alternatives depending on the catalyst used: (1) direct synthesis, (2) via dimethyl ether, and (3) via methanol. Finally, 1,3-butadiene is produced via isomerization/dehydrogenation of n-C4H4 [3]. The effect of process conditions on the feasibility, yield and economics of 1,3-butadiene are studied via steady state simulations in Aspen Plus. Meanwhile, dynamic simulations in Aspen Plus Dynamics are conducted to study the effect of feed fluctuations on the product.

References

1. 1,3 Butadiene (BD) Market Analysis by Application (Butadiene Rubber, ABS, SBR, SB Latex, NBR, Hexamethylenediamine), Bio-based Opportunities and Segment Forecasts To 2020.
2. The Continuing Quest for Butadiene by Jeffrey S. Plotkin.
3. Efficiency Estimation and Improvement of the 1,3-Butadiene Production Process from Lignin via Syngas through Process Simulation Toshiaki Hanaoka, Shinji Fujimoto, and Masaru Yoshida Energy & Fuels 2017 31 (11), 12965-12976.

550024 Estimating the Thermophysical Properties of Black Liquor from Kraft Pulping

Black liquor is a complex mixture that results from the kraft pulping process and is highly toxic and environmentally unsafe. Typically, black liquor is treated after digestion to recover and regenerate the cooking chemicals and liquor. There are three main components of black liquor: lignin, carbohydrate degradation products, and extractives, such as fatty acids and resin. The largest portion of black liquor contains high-molecular weight compounds which precipitate as the liquid is acidified. However, it is the low-molecular weight compounds left in solution that are of most concern when developing thermodynamic models. Even though the low-molecular weight portion of black liquor is minor, there have been hundreds of components identified within the remaining solution. Because of the diversity within the low-molecular weight components, identifying the correct thermophysical properties is crucial for obtaining a rigorous thermodynamic model of black liquor. This work aims to begin the process of identifying the necessary thermophysical properties from liquid-liquid and solid-liquid equilibrium data to initiate the process of building a rigorous thermodynamic model for the kraft pulping process.

549875 Multifunctional Materials for Biomass Conversion to Fuel: A Three-Step Cascade Reaction

Sustainably meeting the growing energy needs of the planet is one of the 21st century's grand challenges. Global energy demand will increase nearly 60% in the next 25 years. Many pathways to meeting our energy needs while reducing fossil-fuel consumption have been posited, all with challenges. One possible route is to convert biomass into fuels.
In this work, we designed a three-step process that includes hydrogenation of phenol, aldol condensation, and hydrogenation of aldol reaction product to produce fuels from oxygenates using palladium containing dendrons supported on ordered mesoporous silica (OMS). Palladium/dendron-OMS catalyzes the phenol hydrogenation and hydrogenation of aldol coupling products while dendron-OMS without Pd was used for the aldol condensation. 85% yield of cyclohexanone was obtained in the first step (phenol hydrogenation). It is also shown that several aldehydes can be used in Aldol reactions with cyclohexanone, including 4-nitrobenzaldehyde (4-NBA), hydroxymethylfurfural (5-HMF) and benzaldehyde (BA). The aldol products were hydrogenated using Pd/dendron-OMS yielding products not dissimilar from what one observes from aqueous phase reforming (APR). The operating conditions of low temperature and low pressure hydrogen in aqueous media is consistent with green chemistry goals.

550072 The Case for Open Innovation

Open Innovation is a term used extensively in companies of all sectors to encourage cooperation between companies, academia and government to reduce risk exposure and leverage capability in developing and commercializing technology. Shell has a decades-long history of open innovation in Shell GameChanger and continues today in support of its new formed New Energies business.

The process of managing open innovation does not follow the typical path in business which have historically developed technology with in-house capability and resources in order to extract as much value as possible. New ways of working and thinking in technology organizations must be adapted to leverage the possibilities of open innovation.

This talk will include generic case studies in which open innovation was successful and led to new products lines, and cases in which open innovation did not consider all business model or supply chain model aspects and failed to produce commercial success.

543895 Open Innovation Experiences: Delivering Success and Avoiding Pitfalls

Open Innovation Experiences: Delivering Success and Avoiding Pitfalls

Open Innovation with a strategic partner can bring your technology to the market quickly and efficiently. However, if not managed well it will lead to project stagnation and missed opportunities. MATRIC has collaborated in over 200 projects with over 100 different customers. The projects have had varying degrees of technical and commercial success, and the learnings highlight key considerations in executing open innovation development opportunities.

I will discuss three projects that used three different approaches to open innovation and how these impacted the project execution and success. The three projects had several similarities:

- the early stage technology discovery work was completed before contacting MATRIC
- the customers wanted to maintain full ownership of the intellectual property
- they didn't have the resources (people and/or facilities) that could be used to develop the technology
- they had external stakeholders in the success of the project who were not engaged directly in the technology development but did contribute time, money or resources to the development effort.

The first customer is a mid-sized company developing a specialty product that would be differentiated in the market by its performance, and could have additional value to the company through the use of lower cost raw materials than their current alternatives. Their open innovation challenge was to reduce the process costs, both operating and capital costs, primarily through an increase in the catalyst productivity. The customer had identified the molecule of interest internally, but did not have the catalysis skills and process development capabilities to prepare for commercialization. Additionally, the target consumers of the product would need to be involved in the final qualification. MATRIC formed a joint technology team with the customer that included the catalyst and process expertise of MATRIC with the product development team from the customer. The team worked very closely together, with weekly meetings and frequent visits by the customer to MATRIC labs to observe the progress and discuss the results. The team was very successful, achieving a 100-fold increase in catalyst productivity well ahead of the project schedule. The consumer of the final product was engaged partially throughout the work, but a critical product test was not completed until late in the technology development. The test results indicated that product needed to be reformulated to avoid adverse effects. This resulted in reworking the catalyst development , with the resulting increased development costs and delayed implementation.

The second customer was interested in developing technology that would create new markets for its product. Earlier work at the laboratory bench scale had demonstrated a technology that could be used to produce an existing commodity product, however the process economics needed to improve. A team was formed of MATRIC experts in process chemistry and process development, along with representatives of the customer and experts familiar with the customer's suppliers and their related technologies. The team had virtual meetings biweekly with periodic face-to-face meetings that included the team members and representatives of the customer's suppliers. The project started with a techno-economic assessment to understand the critical areas for technology development that will lead to project success. It has developed from a bench scale to a larger-scale laboratory study to better understand some of the process challenges and prepare for pilot plant demonstration of the technology. The team communications have allowed the project to progress rapidly, but in a disciplined manner, and to be ready for a smooth transition to the next phase of the technology development.

The final customer was interested in leveraging their technology to create a new path to an existing product for a key consumer. Although their process was a key linchpin in the manufacturing process for the ultimate product, process discovery and development was needed to produce the final product. The customer worked with another research organization to develop the process on a bench-scale, but needed more capabilities to scale the process to a pilot plant and produce large market development samples. The customer shared with MATRIC the reports from the other process development organization and looked to MATRIC to design the pilot plant facility from these reports. Clarifications of the technology were handled by the customer, who would occasionally contact the previous research organization when necessary. This slowed the transfer of the technology and eventually led to a late learning that significantly impacted the process design, requiring redesign of whole sections of the pilot plant with significant increases in the engineering and construction costs. The delays and increases impacted the market development opportunity, and eventually the project was placed on hiatus.

The key learnings from these projects are:

- It is important that in open innovation the customers and research organization must form a partnership with close communication
- All key stakeholders need to be represented in the project to make certain that the process development delivered meets all expectations
- Technology transfer between project stages needs to be thorough. This is readily achieved by having continuity within the process development team, and especially if the open innovation partner is capable of carrying the project through the multiple phases.

557365 Open Innovation Model Successes-New Process in Spray Drying

A success story regarding Open Innovation techniques employed by Pulse Combustion Systems will be presented along with case studies.
We start with a brief background of our technology, followed by a description of our recent focus on solving customer problems via Cooperative R&D. We will present lessons learned and will also show how the company is growing rapidly as a result of using Innovation in a proactive and collaborative manner to solve customer problems. Four cases will be presented to highlight the successes and challenges of the Open Innovation Model.

550567 Use of Process Imaging to Improve the Performance of a Filter

Solid-liquid separation is rate-limiting step for many DOE waste treatment processes. Understanding the fouling structure and its relationship to the foulant may shed light into improving cleaning cycles (minimizing waste streams) and optimizing filter performance (extended lifetime operations). The work used process imaging technologies to quantify dynamics of solid particle movement and filter cake formation on surface of filters, and measure the shape, thickness, dimensions, and buildup of the filter cake as a function of time.

551199 Effect of Different Melt Mixing Methods on the Thermomechanical and Rheological Properties of LLDPE/HNT Nanocomposite

In polymer nanocomposite production good mixing is always desired to attain good material properties. Beside the interactive chemistry of constituents in the mixture, the condition of interface between components and the degree of dispersion strongly influence the performance of final nanocomposite.

Polymeric nanocomposites prepared with halloysite nanotube (HNT) nanofillers have gained strong interest in recent years. HNT is a natural nanotube which is abundantly available at a low price. Compared to other nanoparticles, it has better dispersion due to less inter-tubular contact and its unique geometry.

This experimental research focuses on fabricating nanocomposite with linear low density polyethylene (LLDPE) and HNT with different amounts of HNT loading and using different melt mixing methods. The melt mixing method among all other available techniques is the most compatible with the current industrial compounding and processing technology which has environmentally friendly nature. In this study the effects of different melt mixing methods on the thermomechanical and rheological properties of prepared LLDPE/HNT nanocomposites will be elucidated.

551157 The Effect of Temperature on Flow Field in a Cyclone Separator By Numerical Simulation

The effect of temperature on flow field in a cyclone separator by numerical simulation

Qing Wei[a,b]; Guogang Sun[a,b,*]; Yunnan Jiao[a,b]

[a]*State Key Laboratory of Heavy Oil Processing, China University of Petroleum, Beijing 102249, China*

[b]*Beijing Key Laboratory of Process Fluid Filtration and Separation, Beijing 102249, China*

Corresponding author: Guogang Sun; Tel: +86-10- 89734820;

E-mail:ggsunbj@163.com

Abstract: Temperature is one of the most important parameters for the performance of a cyclone separator. In this study, a computational fluid dynamics is used to predict the flow field in a volute cyclone separator using the Reynolds stress transport model when temperature ranges from 293 K from 1223 K. The numerical results are compared with the experimental data and good agreement is observed. The results show as temperature increases, the tangential velocity and radial velocity of cylindrical and the conical sections decrease, but axial velocity remains almost to be the same. The temperature increase results in a decrease of the tangential velocity along the axial height, and the amplitude of velocity fluctuation. Meanwhile the inner vortex size is reduced at high temperatures. With temperature increasing, natural vortex length and the frequency of vortex end first decreases dramatically and then remains essentially unchanged after temperature above 823K. The effect of temperature on gas flow is achieved by changing the gas viscosity and density. As temperature increases, Reynolds number decreases with the inlet velocity and dimensions of cyclone separator unchanged, which is the root cause of the flow field changing.

Key words: cyclone separator; temperature; flow field

References

- Jianyi Chen, Chunxi Lu, Mingxian Shi. Experimental measurement on flow field in a high-temperature cyclone separator. 2010, 61(9): 2340-3245.
- P A Patterson, R J Munz. Gas and particle flow patterns in cyclones at room and elevated temperatures. Canadian Journal of Chemical Engineering, 2010, 74(2): 213-221.
- M S Shi, H S Kim, D S Jang. A numerical and experimental study on a high efficiency cyclone dust separator for high temperature and pressurized environments. Appl. Therm. Eng. 2006, 25(12): 1821-1835.
- Parker R, Jain R, Calvert S. Particle collection in cyclones at high temperature and high pressure. Environ. Sci. Technol., 1981, 15(4): 451-458.

546646 Dynamic Performance of Flow Field in an Cylindrical Cyclone Separator

This study presents a flow analysis on an atypical cylindrical cyclone separator with a large ratio of length to diameter. Due to the distribution of the flow field determines the merits of separation performance, the aim of this study is to optimize the flow. The Phase Doppler Particle Analyzer (PDPA) is used to measure the gas flow, and the numerical simulation is carried out by using the Reynolds Average Navier-Stokes (RANS) equation, the Reynolds stress model (RSM). Since the model is validated by good agreement between the numerical results and experimental data, and then the tangential velocity is analyzed from the transient flow field and dynamic performance.

The results showed that the tangential velocity of the transient flow field existed a non-axisymmetric phenomenon in the cross section, which was mainly presented as asymmetric distribution of the contours. The zero location of tangential velocity value was not coincide with the geometric center, where the tangential velocity was larger near the zero location. In addition, the dynamic character of tangential velocity was high-speed pulsating, with quasi-periodic characteristic. The structure or operation optimization could change the pulsation frequency and reduce the industrial vibration, which is highly significant.

Meanwhile, a parameter of RMS was used to identify the degree of turbulence velocity deviating from mean velocity. The RMS data showed that the flow instability gradually increased with the fluid flowing into the cyclone separator from the inlet section, until reached a peak value at the position about 2.0 times the length to diameter. And then the flow instability gradually decreased, with the fluid energy dissipated.

551222 Reactor Performance Assessment Using Experimental Studies and Multiscale Modeling: Liquid Phase Ethylene Production By Hydrogenation of Acetylene

Trickle bed and two-phase upflow packed bed reactors have found extensive applications in petroleum, petrochemical, chemical and environmental industries especially for processes involving complex mechanisms. Assessing the reactor performance for these reactors needs extensive experimentation complemented by validation by a reliable mathematical model at different scales. Various factors like incomplete catalyst wetting, liquid maldistribution, interphase mass transfer, intraparticle diffusion are necessary to predict the reactor performance. Moreover, accounting the variation in the overall effectiveness factor across the reactor axis due to the change in the local reactant concentrations is also significant. A one-dimensional mathematical axial dispersion flow model integrated with a pellet scale model has been developed. The kinetics and the related parameters involved in the processes were accounted in the model along with different particle to liquid wetting conditions and correlations. The model was validated using different experimental studies at a range of operating conditions from literature. The hydrogenation of acetylene in the liquid phase for ethylene production will be considered as the model reaction to assess the reactor performance and help in obtaining meaningful scaleup parameters. Gas phase catalytic hydrogenation of acetylene to produce ethylene is commonly practiced in industries to increase the ethylene yield from the cracker. This process has green oil formation which leads to catalyst deactivation and sometimes reactor runaway risks. To overcome this issue as well as to increase the selectivity and conversion liquid phase hydrogenation is under current research. The catalyst life and stability increases by reducing the green oil formation. The solvent (liquid) selectively absorbs acetylene from the cracked effluent, and this liquid phase improves heat transfer in the reactor. The kinetics studies were investigated in a stirred tank reactor in both slurry and basket modes with liquid in batch and gas as continuous using a Pd/Al_2O_3 catalyst . N-Methyl Pyrrolidone (NMP), a polar solvent, was used as the liquid phase. The reaction was conducted at a pressure of 250 psig and temperatures ranging between 60-120°C. In order to estimate the kinetic parameters, a model was developed based on Langmuir-Hinshelwood-Hougen-Watson approach. The dispersion parameter and liquid holdup obtained from the residence time distribution studies by conducting cold flow experiments using a liquid tracer will be integrated into the model. At Multiphase Flows and Reactors Engineering and Applications laboratory (mFReal), a high-pressure, high temperature unit has been built to conduct sensitive reaction experiments. This unit is equipped with a blast shield and has laboratory scale reactors with thermowell and gas-liquid distributors. The unit will be tested for selected operating conditions for the acetylene hydrogenation in the liquid phase to demonstrate the safe experimentation as well as improving the conversion of the process. The experimental data will be validated using the developed multiscale model to assess the reactor performance and further optimize the operating parameters, process time and number of trials.

551349 Minipharm: A Miniaturized Pharmaceutical Manufacturing Platform for Process Development

The traditional approach to pharmaceutical manufacturing is to use consecutively larger batch processes for scale-up; however, there is an alternative that is gaining support: continuous manufacturing. Continuous pharmaceutical manufacturing consists of the integration of multiple unit operations, from synthesis to purification, in a single manufacturing process which can improve scalability, while reducing equipment resource requirements and processing time between unit operations. Small-scale continuous manufacturing would allow companies to implement end-to-end processes that reduce chemical usage, require smaller liquid volume when compared to traditional batches, and have smaller space requirements. The motivation for this paper was to detail the development of an end-to-end miniaturized pharmaceutical manufacturing platform (MiniPharm) with a reconfigurable, modular design to carry out flow reaction, extraction, and crystallization unit operations while utilizing process analytical technology (PAT) tools for molecule identification and quantification. A major advantage for the platform is its ability to simultaneously screen purification routes. Two purification routes following synthesis were compared to determine the impact of unit configuration on product purity. In order to achieve these results, the platform implements two process configurations: synthesis, followed by nitrogen-segmented anti-solvent crystallization and synthesis followed by inline extraction, separation, and nitrogen-segmented crystallization. Nitrogen-segmented crystallization is of particular interest because it allows for the segments to act as individual crystallizers. The aforementioned approach to continuous crystallization has not been demonstrated in an integrated design with continuous reaction and flow extraction unit operations.

The compound selected in this study was Diphenhydramine hydrochloride (DPH). The reaction consists of a single step and has been shown as an example of atom economy due to the direct synthesis of the final product without the necessity of an extraction/solvent swap step. The reaction provided an opportunity for development of the platform with a molecule that is suitable for continuous small-scale production. Platform development included individual unit operation characterization and the coupled unit operation screening prior to the complete design integration. Initial unit operation studies provided valuable insight into system parameters necessary for system integration.

The implementation of PAT was critical from the start of platform development. Monitoring of the module performance was achieved with the use of Ultra high performance liquid chromatography (UPLC) and Mass Spectrometry. Characterization of crystal products was carried out using powder x-ray diffraction (XRD) and Raman spectroscopy. Imaging of the crystallization process in segmented flow was monitored by using an optical camera. Crystal morphology and size distribution data was measured using Malvern Morphologi G3. Reactor module development consisted of a screening multiple residences at various temperatures to determine product concentration. Initial extraction development focused on pH adjustment for product recovery; however, Thin Layer Chromatography (TLC) results provided insight into a non-pH adjusted approach to isolate DPH. Crystallization module development began with cooling only crystallization, but was expanded to include a combined anti-solvent and cooling approach to allow for enhanced product recovery.

In this study, the development of an integrated, reconfigurable platform was detailed for Diphenhydramine hydrochloride process development. PAT tools were used to determine the performance of unit operations in the integrated system. Finally, products from the simultaneous purification routes were compared to determine the effectiveness of additional purification steps on the final product purity. Implementation of the process development framework in a multi-step reaction will be explored in future studies.

551228 Predicting Phase Behavior and Intuitive Solvent Selection Using Mosced

MOSCED (Modified Separation of Cohesive Energy Density) is a powerful analytic tool for predicting phase behavior. Additionally, as a solubility parameter based method, it can be used to gain insight into the underlying molecular-level interactions for intuitive solvent selection and formulation. The major advancement of MOSCED over other solubility parameter methods is its improved treatment of association interactions. Results will be shown for predicting binary phase equilibrium for select systems. Additionally, simple expressions will be presented to use MOSCED to design solvent blends to achieve solubility enhancement and to screen entrainers for extractive distillation processes.

A shortcoming of MOSCED for design applications is the availability of parameters. Results will be shown of recent efforts to expand the parameter matrix, including novel predictive methods. Methods will be demonstrated to predict parameters including LSER (linear solvation energy relationship) parameters, group contribution methods, electronic structure calculations, and molecular simulation.

551169 Advanced Self-Heat Recuperative Distillation System

In the last several years the authors have been developing the self-heat recuperation technology for the heating and cooling thermal processes in which the heat of an effluent stream is recuperated and recycled for feed stream heating by the adiabatic compression of gas and/or vapor. This SHR technology has been applied for the heat circulation of various thermal processes such as distillation, drying, CO2 chemical absorption, cryogenic air separation, pressure swing adsorption, sea water desalination, etc., leading to a considerable amount of energy saving. In the case of the distillation systems the overhead vapor from is pressurized by compressors and its temperature is increased by adiabatic compression. The duties of reboiler and feed heater are supplied by the heat of the discharged streams from the compressors. This results that the heat of the process stream is circulated without the need for any additional heat, and, thus, the energy consumption of a process can be greatly reduced. However, if the boiling temperature difference between two components is large, the energy consumption of distillation becomes large because the minimum work required for the heat circulation by the self-heat recuperation is almost proportional to the temperature increase by the adiabatic compression.

In the present paper, a novel self-heat recuperative distillation system with a large temperature difference between top and bottom of column is described based on the exergy and exergy analysis.

550660 Best Practices for Managing Safety during the Design Phase of Pilot Plants

In this presentation, we will cover a range of best practices for managing safety in pilot plants during the design phase. We will also specifically share some of the safety processes/tools that we use at Shell Oil Company including risk assessment matrix, management of change, hazard analysis etc.

1 **Using Ventilation to Reduce the Area Electrical Classification of Pilot Plants and Laboratories**

2 **Introduction**

- Most codes allow use of ventilation to reduce the area electrical classification
- This is an important factor in pilot plant and laboratory installations
- However, there remain many areas that are difficult to determine how to most appropriately apply these codes for safe installations in pilot plants and laboratories

3 **Code Summary**

- NFPA 70 The National Electric Code
  - 500.5(B)(2) Allows a Class I Division 2 classification rather than Division 1 where ignitable mixtures are "normally prevented by positive mechanical ventilation"
  - 500.5(B)(3) allows eliminating a Division 2 area adjacent to a Division 1 area where "communication is prevented by adequate positive-pressure ventilation"
  - But gives no guidance on what is "adequate ventilation"
- NFPA 30 The Flammable and Combustible Liquids Code
  - 17.11.1 Defines adequate ventilation as "a rate sufficient to maintain the concentration of vapors within the area at or below 25 percent of the lower flammable limit (LFL)."
  - 17.11. 2 and 17.11.3 allow the user to determine this by
    - +Fugitive emissions calculation (defined in API 500),
    - +Sampling, or
    - +Providing 1 CFM/SF floor area
  -
-

4 **Code Summary (cont'd)**

- NFPA 497 Classification of Flammable Liquids, Gases, or Vapors and of Hazardous (Classified) Locations for Electrical Installations in Chemical Process Areas
  - Allows a general purpose classification with adequate ventilation if ignitable mixtures "are contained within suitable, well maintained, closed piping systems"; or
  - 3.3.1 Defines adequate ventilation as
    - +6 ACH,
    - +1 CFM/SF, or
    - +Any "other similar criterion that prevents the accumulation of significant quantities of vapor-air concentrations from exceeding 25 percent of the lower flammable limit (LFL)."
- API 500 Recommended Practice for Classification of Locations for Electrical Installations at Petroleum Facilities Classified as Class I, Division 1 and Division 2
  - Repeats the NFPA 497 recommendations and gives a suggested calculation method for "fugitive emission" (AKA leak rate) in appendix B
  - Considers some locations to have inherently adequate ventilation

+Outside

+Outdoor enclosure with only a roof and one side (or equivalent suitably designed windbreaks)

+Indoor enclosure with mechanical ventilation and adequate safeguards against ventilation failure

–

5 **Laboratories**

- NFPA 45 Fire Protection for Laboratories Using Chemicals
  - –5.6.2 "Laboratory work areas, laboratory units, and chemical fume hood interiors shall be considered as unclassified electrically with respect to Article 500 of NFPA 70 unless operations are determined to cause a hazardous atmosphere."
- This second part of 5.6.2 as well as the accompanying annex information is often ignored
  - –"A qualified design professional and owner safety officer should review the laboratory conditions through a hazard analysis and/or risk assessment to determine if a hazardous (ignitable) atmosphere could be developed within the laboratory work area, laboratory unit, and/or fume hood. If a hazardous atmosphere could be developed, these areas should be electrically classified per NFPA 70, Article 500."
  - –Fortunately few laboratories conduct operations in such a manner as to release significant quantities of ignitable mixtures into the laboratory itself
- 45: 5.6.2 is often incorrectly applied to pilot plants although specifically prohibited by 45: 1.1.1(2)
  - –And the presence of pilot plants is often the most credible reason for having to electrically classify a laboratory

6 **Pilot Plants?**

- NFPA 45 does not apply and <u>cannot</u> be used for pilot plants [1.1.3(2)]
  - –You cannot use NFPA 45 to justify a general purpose area electrical classification for the pilot plant
- Pilot plants can be placed in a laboratory which is covered by NFPA 45
  - –In this case the pilot plants are not covered but must be included in any NFPA 45 inventory
  - –You cannot apply any other provision of NFPA 45 to the pilot plant, only to the laboratory

7 **Providing A Fire Wall Would Be A Better Long Term Approach**

8 **How About Hoods?**

- Hoods are generally considered to be so well ventilated that they meet the criteria for keeping any leak below 25% of the LEL
- However, this may not always be the case
- Leaks are drawn to the rear of the hood to the exhaust plenum
- So ignition sources may be encountered before the leak is adequately diluted
-

•
•
•
•
•
•

- Numerous fires and explosions have occurred in hoods over the years
  - Most, fortunately have been small and resulted in only equipment damage
  - Many, have avoided personnel injury more by fortune than mitigative measures
- This is an area that deserves much more attention than it usually receives

9 **Pilot Plant Areas**

- Most organizations with pilot plant areas do one of the following:
  - Provide 6 air changes/hour;
  - Provide 1 CFM/ft^2 or higher;
  - Place the equipment in a hood or ventilated enclosure;
    - +Both of which usually exceed both of the above
  - Provide the API suggested ventilation rate;
    - +Which results in a very low ventilation rate
  - Provide continuous fixed gas detection;
    - +Which is allowed by the codes providing the operations without the detection would only have resulted in a Class I Division 2 classification
      - oThe codes do not allow you to declassify a Class I Division 1 location to unclassified with a gas detection system
  - Classify the area as Class I Division 1 or Division 2; or
    - +Either of which is expensive and limiting
  - Some combination of the above

10 **API 500 Fugitive Emission Calculation**

- Steady state leak

$$Minimum\ Ventialtion\ Rate(CFM) = \frac{Maximum\ Leak\ Rate\ (CFM)}{25\%\ (LEL\ of\ the\ gas)} \text{ X Safety Factor}$$

- The safety factor is usually any where from 2 to 5 and represents the non-ideality of the ventilation
- This produces low ventilation rates
- Example:
  - Hydrogen gas (LEL =4%)
  - 1 LPM (0.035 CFM) leak rate
  - Safety factor of 5
  - $Minium\ Ventilation\ rate = \frac{.035(5)}{.25(.04)}$ = 17.7 CFM
  - Which is less than 0.2 CFM/SF or 1.3 ACH for even a 100 SF area with an 8 ft ceiling
    - +Which is unlikely to do much good
- The problem is the API calculation is based on large leaks in small areas (typical for a

plant)

–At 10 CFM, 5,000 CFM is required which would be 50 CFM/SF or 375 ACH

11 **Concerns**

- The tacit assumption behind all the codes is that the leak/release is diluted below the LEL before it can find an ignition source
- Overall exhaust rates (CFM, CFM/SF, or air changes/hour (ACH)) are often badly miss leading
  - –Higher pressures can result in a concentrated leak shooting significant distances
  - –Many pilot plants have ignition sources very close to potential leak points
    - +Heated reactors
    - +Instrumentation
    - +Heat tracing
  - –Many ventilation schemes are poorly designed
    - +Improper distribution
    - +Short circuiting and failing to sweep the area
    - +Lacking adequate alarms and controls in the event of a ventilation failure
  - –Many calculations use unrealistic assumptions
    - +A uniform, spherical dispersion
    - +No impediments to dispersion
    - +No pressure effects
- And, of course, predicting the maximum leak rate requires a very clear crystal ball
  - –So it is often way more than you assumed

12 **Hoods and Ventilated Enclosures**

- Many organizations install large walk in hoods around their pilot plants believing that these provide adequate ventilation to be an unclassified area
- Common problems
  - –Inadequate face velocity over large portions of the hood face resulting in no to inadequate containment
    - +So the leaks could escape into the unclassified exterior
  - –High exhaust rates near the exhaust inlet but with long distances to these exhausts
    - +So a leak is much more likely to find an ignition source before dilution
  - –Significant dead zones inside the enclosure due to this poor distribution, large solid objects, or high density of components
    - +Raising the potential for build up of ignitable concentrations
- These type of hoods and ventilated enclosures are
  - –Expensive to install
  - –Very expensive to operate
  - –Consume a lot of valuable floor space
  - –Usually provide no more mitigation than the same ventilation properly applied over the entire room
  - –Are often more of a placebo than an effective mitigative measure

13 **Recommendations and Best Practices**

- If you intend to use ventilation to reduce your area electrical classification then it must
  - Be properly designed suggests
  - Be installed so as to sweep the area thoroughly
  - Take into account the reduced effectiveness of large amounts of equipment in the area
  - Probably require more than 6 ACH(10-12 ACH is more likely) and certainly more than 1 CFM/SF (3 CFM/SF is more common)
  - Be reliable enough to work all the time baring a major problem outside your control
    - +E.g. power failures
  - Have a well planned and diligently implemented PM Program
    - +To reduce the potential for equipment failures
  - Have appropriate, fail safe, independent, alarms to warn of a failure
  - Have well conceived, practiced, and credible emergency procedures covering a ventilation failure
- Simply installing a hood may not be enough

14 **Recommendations and Best Practices (cont'd)**

- Carefully evaluate the most credible sources of leaks as part of your hazard analysis and risk assessment
  - Many will likely require additional mitigative measures
- Provide local exhaust ventilation for areas of
  - Higher risk
    - +Reactor seals near electric heaters
    - +Seals on heated and cooled systems
  - Manual operations
    - +Filling
    - +Emptying
    - +Cleaning
  - Sampling
  -

15 **Recommendations and Best Practices (cont'd)**

- Provide positionable local exhausts where the operators may need it for troubleshooting or maintenance
  - Make sure there are enough
  - Make sure they can reach <u>everywhere</u> they might be needed
  - Make sure they can be easily positioned and will stay where positioned for proper protection
  - Make sure the operators know how to use them properly
  - Make sure they are using them properly
- Providing exhaust arms, while expensive, greatly improves operability and effectiveness

–They stay where they are positioned
–They don't need the operator to hold them or wedge them somehow

16 **Recommendations and Best Practices (cont'd)**

- Local enclosures, while theoretically improving exhaust effectiveness, often are simply more costly, more space intensive, and less effective
  - –Usually due to poor design and lack of a plenum

17 **Final Thoughts**

- Ventilation is an incredibly effective mitigative measure to protect against fire and explosion in laboratories, pilot plants area, and other research areas
- It can be used to reduce the required area electrical classification without added risk
- But to do so, it needs to be:
  - –Integrated into the overall area safety design
  - –Thoroughly and fully evaluated as part of the hazard analysis and risk assessment
  - –Properly designed
  - –Properly maintained
  - –Properly used

551680 Safety Practices for Testing Pilot Plants – What to Avoid

This presentation will focus on testing of pilot plants prior to putting them into operation. Incoming inspections of new materials for testing and simulation will be addressed, as well as their safe handling and disposal. The benefit of a safety checklist will be explained, and how it is used to prevent some of the common safety issues that arise during testing. This presentation will deal with what to avoid during testing, based on lessons learned over the years.

551260 Investigating Liquid Mixture Adsorption Equilibria Using Adsorption NRTL Model

Several organic contaminants such as benzene and its derivatives and inorganic contaminants such as heavy metals are discharged into the water sources which has led to widespread contamination of drinking water. Reliable thermodynamic models are needed for the proper design and optimization of adsorption processes which could help in effective removal of these contaminants. Therefore, we explore the applicability of the recently developed adsorption non-random two-liquid (aNRTL) model for predicting the binary and multicomponent liquid adsorption behavior. The model takes into consideration the adsorbate-adsorbent interactions and correctly predicts the negative deviations from ideality for various binary liquid adsorption mixtures. With only a single binary interaction parameter per adsorbate-adsorbate pair, the model can successfully correlate various binary liquid adsorption systems on several kinds of adsorbents. The model should be a simple yet robust engineering tool for the proper design and operation of an adsorption system by accurately predicting the multicomponent adsorption behavior.

550235 *In Situ* biobutanol Recovery from Culture Media By Modified Activated Carbon: Equilibrium, Kinetics, and Reusability

In this research, a modified activated carbon was used to separate 1-butanol (2.35 g/L) produced by genetic modified *Synechocystis*PCC 6803 from culture media. Biobutanol produced from cyanobacteria was considered as a promising sustainable energy candidate for relieving the current energy crisis. The biobutanol can be separated by distillation, liquid extraction, sorption, pervaporation, and so on. The most widely used traditional method to separate the biobutanol from its culture media was distillation. However, with the consideration of the technical difficulties and energy consumption generated from other separation methods to recover this relatively low concentration biobutanol, the sorption that has been neglected on biobutanol recovery can be an alternative sustainable way to reduce the process energy consuming and corresponding minimize the operation cost comparing to the distillation process. To achieve such effective sorption on this low concentration biobutanol recovery without any previous research done before, the normal carbon was experienced a two-step modification to possess a super high hydrophobicity and lipophilicity.

The sorption step in this *in situ*biobutanol separation was simply applied after separating the culture media from the *Synechnocystis*PCC 6803 cells. Without breaking the cyanobacteria cell, the biobutanol, an extracellular product existed in the separated culture media, was directly sorbed by immersing the modified activated carbon into the media for 24 hours. The modified activated carbon after 24 hours sorption went through a rough filtration process by pouring the liquid media and modified activated carbon through a mesh. The resulted modified activated carbon was sent to the distillation for the desorption of the sorb liquid. Then, a layer separation was observed in the desorption liquid, with a top layer mainly consisting of butanol and a bottom layer mainly consisting of water according to the gravity. This layer separate approved an irreplaceable successful in both energy and cost saving of the sorption process compared to those of the distillation. Since in distilling the biobutanol with a low concentration of 2.35g/L, a layer separation can be found after distilling the biobutanol for the first time. Hence, for separating the same amount of biobutanol, the sorption cooperated with the distillation can significantly reduce the liquid amount that needed to be distilled. The liquid amount that needed to be distilled already concentrated on the biobutanol concentration by experiencing the sorption procedure. Meanwhile, the butanol concentration left in the culture media after sorption was as low as the butanol concentration after about 20 hours distillation, which also exhibited the energy saving aspect of the sorption procedure. Compared to other biobutanol recovery technology, such sorption technique that has been neglected was demonstrated highly effective in both energy consumption and cost expense. For the estimation of the operation cost and energy efficiency, the reusability of the modified activated carbon was also studied. The corresponding results of the reusability experiments strengthened the above conclusion by testifying the sorption capacity of the as-prepared hydrophobic activated carbon for 1 month reuse studies. To understand the sorption mechanisms better, both sorption kinetics and equilibrium was investigated on this biobutanol recovery procedure by comparing different mathematic models.

544151 High Temperature Hydrogen Attack - What Is It? What Is Industry Doing about It?

High temperature hydrogen attack (HTHA) is a damage mechanism most commonly occurring in carbon steel and C-0.5Mo steels that are exposed to hot hydrogen service. The HTHA damage mechanism and historical operating limits (Nelson Curves) for HTHA will be reviewed. Inspection for the HTHA has proven to be challenging as it is a localized damage mechanism that can manifest as either tight cracks or diffuse volumetric damage. Unfortunately, historic design and historic inspection methods have been relatively ineffective.

Assessing HTHA risk can be controversial and has resulted an Industry push for better understanding of the damage mechanism and a shift away from historic operating limit. This resulted in a significant gain in understanding of the damage mechanism and emerging inspection technologies within recent years. Select case studies will be presented and discussed. The direction and steps industry is taking to successfully address HTHA risk and develop tools for integrity programs will be reviewed.

550614 Laboratory Safety for the Novice Lab Technician

A recent college graduate is faced with significant and new safety concerns when hired to work in a commercial petrochemical testing laboratory. The pace of testing, the potency of reagents, and the breadth of equipment is often a dramatic shift from the campus "O-Chem Lab" she experienced while pursuing her degree. In this talk, we will discuss laboratory best practices and how the onboarding process for a new hire can establish a culture of safety within the lab.

550662 Best Practices for Managing Safety during the Commissioning and Start-up Phases of Pilot Plants

This presentation will cover a range of best practices for managing safety in pilot plants during the commissioning and start-up phases. We will also describe issues related to process and personnel safety from multiple pilot plants from process development projects at Shell Oil Company.

PROCESSING WITH SOLIDS IN A LIQUID-LIQUID EXTRACTION COLUMN

Authors:

Donald J. Glatz
Brendan Cross
Thomas Lightfoot

Koch Modular Process Systems, LLC

Prepared for Presentation at

AICHE Spring Meeting
New Orleans, LA
April 2019

Introduction

The typical liquid-liquid extraction process involves counter current contact of two immiscible liquids in order to affect mass transfer of a specific component or components from one liquid phase into the other. In some instances, the presence of solids can complicate the separation process. Most extraction columns cannot tolerate insoluble solids, primarily due to material building up on the internals and/or slow settling velocity for the solids. In some instances, the solute to be removed is within the solid particles. For these processes, the extraction column must be capable of improving the mass transfer out of the solid particles, while still providing sufficient settling time to allow for counter current flow. The KARR® Reciprocating Plate Column has been proven to handle many different types of slurry feed processes, including particulates, crystals, and powders.

This paper will present three cases where pilot plant testing in a KARR® Column was used to demonstrate the feasibility of extraction processes with significant amounts of solids present and to develop the data required for scale up to commercial columns. An additional problem with pilot testing is providing a method for feeding and controlling the flow of slurries on a very small scale as well as handling the bottom outlet of the extraction column. For each of the cases presented, unique test methods were developed in the pilot plant, with individualized equipment set-ups.

KARR® Column – Best Choice for Handling Solids

There are many different types of column available for liquid-liquid extraction (1). However, when solids are present, most extraction columns are not acceptable for long term processing. A summary of the different columns follows:

Static Column (Packed Column or Sieve Tray Columns)

Solids can easily build up in packed columns using either structured packing or random packing because the internals are stationary. Once solids build up on the internals the coalescing behavior can change resulting in loss of efficiency. When a significant amount of solids build up, the capacity can be reduced resulting in premature flooding. Sieve tray columns also have the stationary trays with a lot of horizontal surface area for the solids to build up.

Columns with Rotational Agitation

All columns with rotational agitation (such are RDC, SCHEIBEL® and Kühni Columns) have stationary baffles or trays. These horizontal surfaces provided "dead zones" where the solids can build up. Although better than a static column, these columns also have difficulty handling any significant amount of solids due to build up on these surfaces.

Pulsed Columns

Pulsed Columns have no moving internals. The internals are stationary and the agitation is provided by pulsation of the liquids. Like a static column, the stationary internals provided surfaces where the solids can build up.

The KARR® Column is unique in that the entire plate stack reciprocates, i.e. there are no stationary internals. Thus, there are no dead zones nor stationary plates where solids can build up. The holes size on the perforated plates is typically 9/16" diameter, which is large enough to handle relatively large particulates. As can be seen in the case studies presented next, the KARR® Column will handles slurries with up to 40% or more suspended solids and allow for long term processing without solids buildup.

<u>**Case Study #1 – Extraction of Salts from Polymer Crystals**</u>

The application involved the removal of salts from a feed slurry containing approximately 94% solvent (methanol and methyl acetate) and 6% polymer. The polymer was in the form of crystals of many shapes and sizes, with the largest crystals approximately 1/8" long.

A number of different equipment set-ups were attempted when studying this process. In the end, the best functioning set-up is shown in Figure 1. The column was a 2" diameter, glass shell KARR® Column with 316SS plates and spacers. Significant other features of the set-up are described as follows:

- The slurry feed system consisted of an agitated drum, a circulation loop, and a solenoid valve at the top to meter slurry into the top of the column. A centrifugal pump was used to circulate the slurry from the bottom of the drum, through a 1" PVC pipe that looped above the column feed location and back into the drum. A 2" Tee was installed at the top of the loop which led to a 2" air actuated solenoid valve. The valve was controlled by a timer that cycled ON/OFF to allow intermittent feeding of the slurry into the top of the column. A 1" vacuum breaker line was installed into the circulation loop immediately downstream of the solenoid Tee. This eliminated the vacuum pull on the top of the column and allowed smooth feeding during the test.
- A pump supplied by Fluid Metering Inc. (FMI) was used for metering the solvent to the bottom of the column and a gear pump was used for filling the bottom chamber after each time the solids were discharged.
- The bottom chamber included 2 x 3" ball valves with a 3' x 36" glass collection chamber in-between. A solvent inlet nozzle and vent line were installed directly below the upper ball valve in the glass section. When the bottom chamber was filled with solids, the top valve was shut, vent line opened and then the bottom valve opened to discharge the solids into a 5-gallon pail. After closing the bottom 3" valve, solvent was pumped in to fill the bottom chamber. When full, the vent was closed and upper 3" ball valve opened to allow solids to build up until the next cycle.

Prior to actual testing the slurry feed rate was calibrated. The circulation pump was set at a constant speed and the slurry rate determined for a number of different ON/OFF cycles of the solenoid valve. This resulted in a calibration curve for determining the slurry feed rate. For testing, the column was initially filled with solvent and the agitation set at the desired speed (strokes per minute). While the column was filling with solvent, the circulation pump and drum mixer were on. When the column was full, the solenoid valve was turned on and set at the desired ON/OFF cycle. The solids were observed to flow smoothly down the KARR® Column with no build-up on the moving perforated plates. Solids were allowed to build up in the bottom chamber until about 90% full. At this point the top valve was closed and the discharge procedure described above was followed. This procedure was repeated as required throughout the run and then the final sample of solids was collected from the bottom chamber and analyzed.

Results and Conclusion

The test successfully demonstrated successful operation of the KARR® Column. Solids were observed to flow freely down through the up-flowing solvent phase with no build during successful operation. Mixing by the reciprocating plates provided sufficient hold up and turbulence for the solid particulates. All runs using pure methanol as the solvent generated the required salt reduction. Sufficient data was generated using pure methanol as the solvent to design the production column which could be offered with a process performance guarantee.

Case Study #2 - Extraction of Oil from Corn Meal

The application involved the extraction of oil from ground corn meal solids that contained approximately 4% corn oil. For the process, the corn meal solids were to be initially slurried into ethyl acetate and fed to the top of a column with fresh ethyl acetate fed to the bottom of the column.

The equipment set-up for this process is shown in Figure 2 and included a 2" diameter x 10' plate stack height KARR® Column with glass shell and 316SS plates and spacers. The agitation zone of the column was wrapped with electrical heating tape to maintain the temperature in the column. Additional features of the set-up included:

- Initial attempts to feed the corn meal as a slurry in ethyl acetate proved to be unsuccessful due to equipment limitations. Instead, a rotary valve was used for feeding dry corn meal solids or wet corn meal cake (wetted with ethyl acetate) into the top of the column as shown in Figure 2. A stainless steel funnel was attached to the top of the rotary valve. The only successful method for feeding the corn meal solids was determine to be by manually "sprinkling" the corn meal into the throat of the rotary valve, just fast enough to keep the vanes full.
- An FMI pump was used for metering ethyl acetate through a mass flow meter and into the bottom of the column. A heat exchanger was included in the solvent feed line for preheating the solvent prior to entering the column.
- The bottom of the column included a 2" ball valve and a 2" rotary valve in series. It was found that the rotary valve alone could not maintain the solid-liquid interface due to excessive seepage of the liquid through the valve. Therefore, the rotary valve was left on all the time and the solids level maintained between a certain range in the bottom chamber by manually opening and closing the ball valve, allowing the wet solids to discharge into a drum.

During testing, the column was initially filled with ethyl acetate, the agitator turned on and agitation speed set. While the column was filling, the heat exchanger and heat tapes were adjusted to achieve the required temperature in the column. When the column was full, the corn meal solids were manually metered into the column at the specified rate by adjusting the RPM on the rotary valve. A 250 ml beaker was utilized by the operator to pour the corn meal into the throat of the rotary valve. Therefore, the feed rate was determined by how often a beaker of corn meal was charged to the column. During testing, the corn meal solids flowed smoothly down the KARR® Column with no build up on the internals. Solids were allowed to build up in the glass section at the bottom of the column. While filling the bottom rotary valve was left on and the bottom ball valve closed. When the desired solids level was achieved, the ball valve was opened until most of the solids discharged from the column and then the ball valve closed again until the next cycle.

Results and Conclusions

The KARR® Column was determined to be feasible for counter current contact of ethyl acetate and ground corn meal as either a dry solid or as a pre-mixture in ethyl acetate. A small amount of fines did carry over with the ethyl acetate extract phase (< 1% total). The target corn oil recovery was achieved. Pretreatment of the corn meal with ethyl acetate to create a "wetted" solids feed resulted in greater corn oil recovery. Data was available for scale up to production size columns. For this process, a demonstration plant was constructed including an 8" diameter, KARR® Column for counter current processing the corn meal with ethyl acetate. This column was reported to operate successfully, achieving the specified performance requirements.

Case Study #3 – Fractional Extraction with Middle Slurry Feed

The application involved a unique fractional extraction process (2) in which water was fed to the top of a column and an immiscible solvent was fed to the bottom of the column. The middle feed was a slurry in the organic solvent containing 5-10% of fine powdery solids. The purpose of the extraction process was for the solids (product) to dissolve into the water phase and phenolic compounds to extract and leave with the solvent phase out the top of the column.

The set-up for testing this process is shown in Figure 3 and included a 1" diameter x 10 plate stack height KARR® Column with glass shell and 316SS plates and spacers. Additional features for the set-up are as follows:

- The slurry feed was made in a 35-gallon, jacketed, stainless steel reactor vessel with four internal baffles and equipped with a 1 HP agitator. The agitator had a 6-blade Rushton turbine impeller. Hot oil was circulated through the jacket to maintain temperature.
- Variable plate spacing was used in the KARR® Column. At the mid-point of the column there was 6" plate spacing, with the slurry inlet location centered on this 6" gap. Below that there were two plates spaced 3" apart and then standard 2" spacing to the bottom of the plate stack. Above the slurry inlet, there were 2 plates with 4" spacing, followed by 3 plates with 3" spacing and then standard 2" spacing to the top of the plate stack. Greater spacing was used above and below the feed point to inhibit flooding conditions because of the greater amount of mass transfer in this section of the column.
- FMI pumps were used for metering the organic solvent and water through mass flow meters and into the column. For all runs the water entered above the top plate and the solvent below the bottom plate. A gear pump was used to circulate the slurry out of the bottom of the reactor and back into the top. The circulation line included a tee with a feed line to the middle of the extraction column using an FMI pump to regulate the flow rate.

- An FMI pump was used to discharge the heavy phase from the bottom of the column and to manually control the interface in the upper disengaging chamber. The light organic phase was allowed to overflow from the column into a receiver.

Initially the column was filled with DI water (setting water as the continuous phase) until the water level reached the top of the column. When the column was full, the water and solvent rates were set as required and the initial agitation speed set. The interface was established and maintained in the top disengaging chamber. Next the slurry feed pump was turned on and the feed rate to the middle of the column established. For each run, the agitation speed was set at the desired strokes per minute for the test. The fine solids were observed to enter the column and immediately become well dispersed. In the middle of the column, the solids were observed to flow easily through the perforated plates with no build up on the plates. All the solids flowed down with the aqueous phase and were fully dissolved within 12" of the inlet location.

Results and Conclusions

The pilot plant testing clearly demonstrated the feasibility of using the KARR® Column for successfully processing this fractional extraction process with a slurry feed to the middle of the column. As long as sufficient amount of water was used (to solubilize the solids) and the temperature was maintained above 75 °C, the solids were observed to dissolve in the water phase well before the bottom of the column. Essentially 100% of the solids material was recovered in the aqueous phase while >99.5% of the phenolic impurities were removed with the solvent phase. Optimized data was generated during pilot plant testing to scale up and design a production column which could be offered with a process performance guarantee. A large KARR® Column was subsequently designed and fabricated. Start-up was still pending at the time of this paper.

Closing Remarks

There are some applications for a liquid-liquid extraction process for which solids may be present in the feed. The KARR® Column is the ideal column type for these processes, primarily because the entire plate stack is in constant motion, thus not allowing any stationary surfaces where the solids can begin to accumulate and build up. Here we have demonstrated three different types of application for which the KARR® Column was successfully tested and data generated for scale up to production size equipment.

Literature Cited

1. Cusack, R., Fremeaux,P., "A Fresh Look at Liquid-Liquid Extraction, Inside the Extractor", Chemical Engineering, March 1991

2. Glatz, D., Lightfoot, T., "Designing Columns for Fractional Liquid-Liquid Extraction", AIChE National Meeting, November 2010.

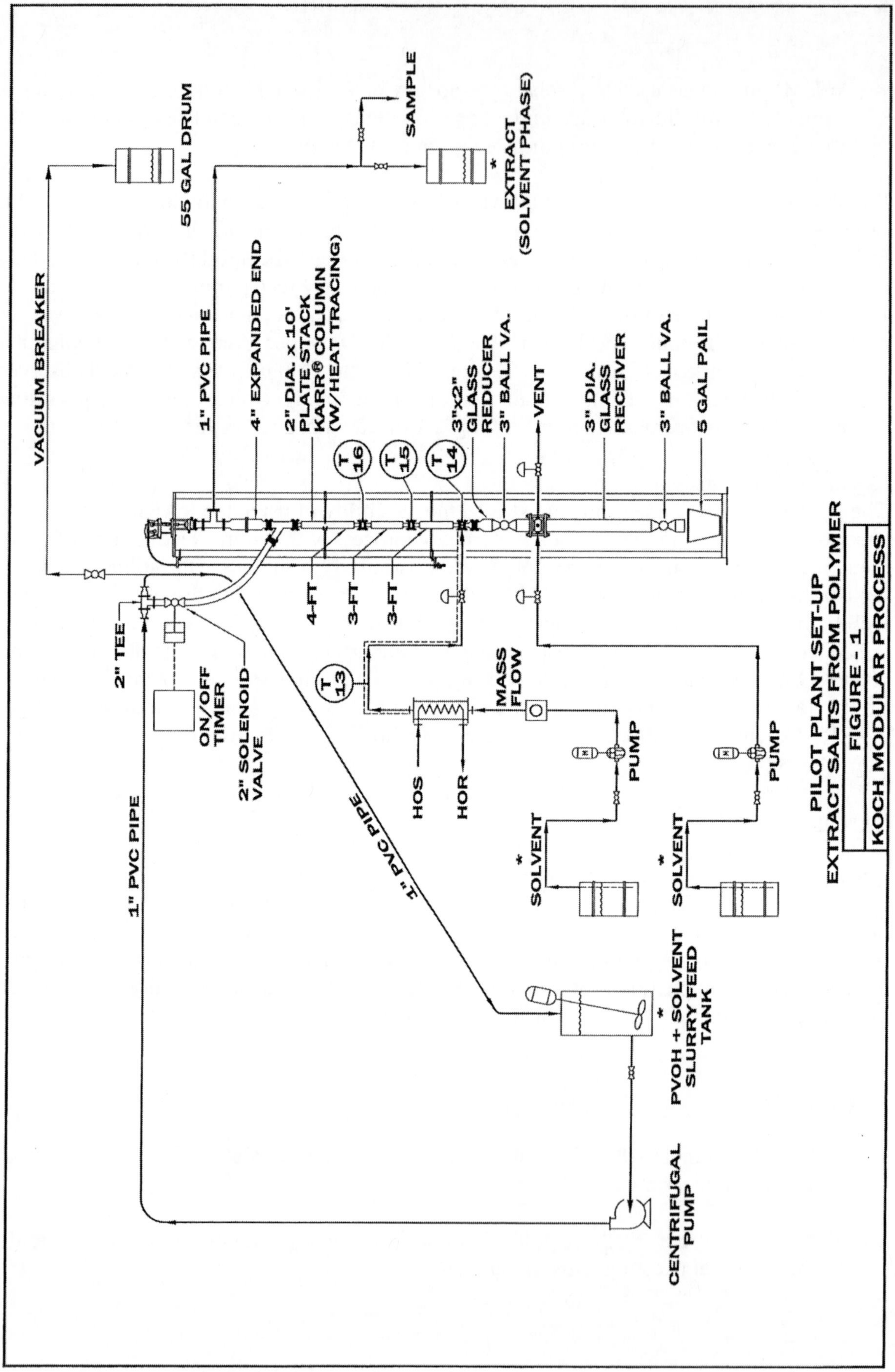
VACUUM BREAKER
55 GAL DRUM
1" PVC PIPE
4" EXPANDED END
2" DIA. x 10' PLATE STACK KARR® COLUMN (W/HEAT TRACING)
SAMPLE
*
EXTRACT (SOLVENT PHASE)
T 16
T 15
T 14
3"x2" GLASS REDUCER
3" BALL VA.
VENT
3" DIA. GLASS RECEIVER
3" BALL VA.
5 GAL PAIL
4-FT
3-FT
3-FT
2" TEE
ON/OFF TIMER
2" SOLENOID VALVE
T 13
MASS FLOW
HOS
HOR
PUMP
PUMP
* SOLVENT
* SOLVENT
1" PVC PIPE
1" PVC PIPE
* PVOH + SOLVENT SLURRY FEED TANK
CENTRIFUGAL PUMP
PILOT PLANT SET-UP
EXTRACT SALTS FROM POLYMER
FIGURE - 1
KOCH MODULAR PROCESS

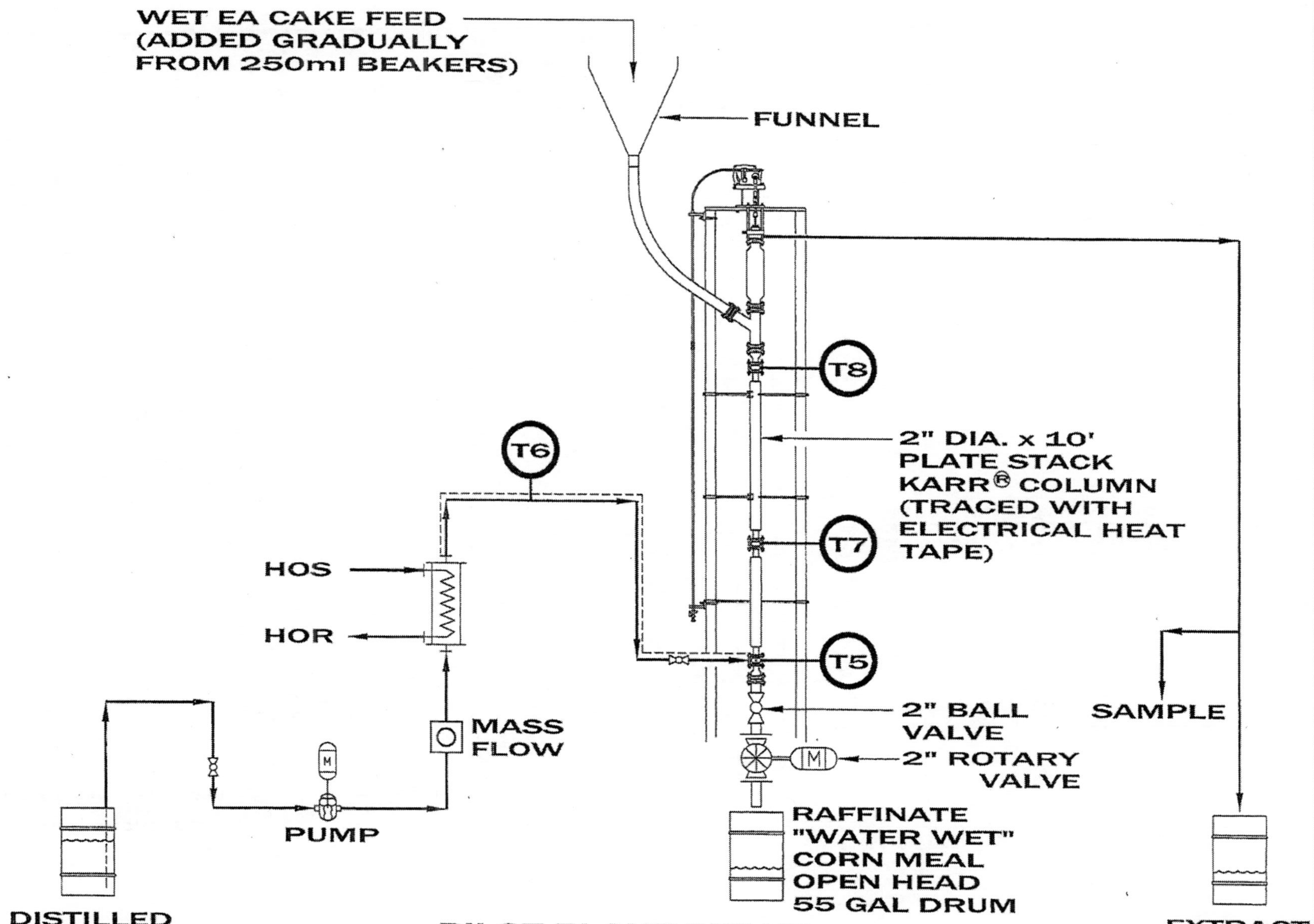

PILOT PLANT SET-UP
WATER EXTRACTOR

FIGURE - 2
KOCH MODULAR PROCESS

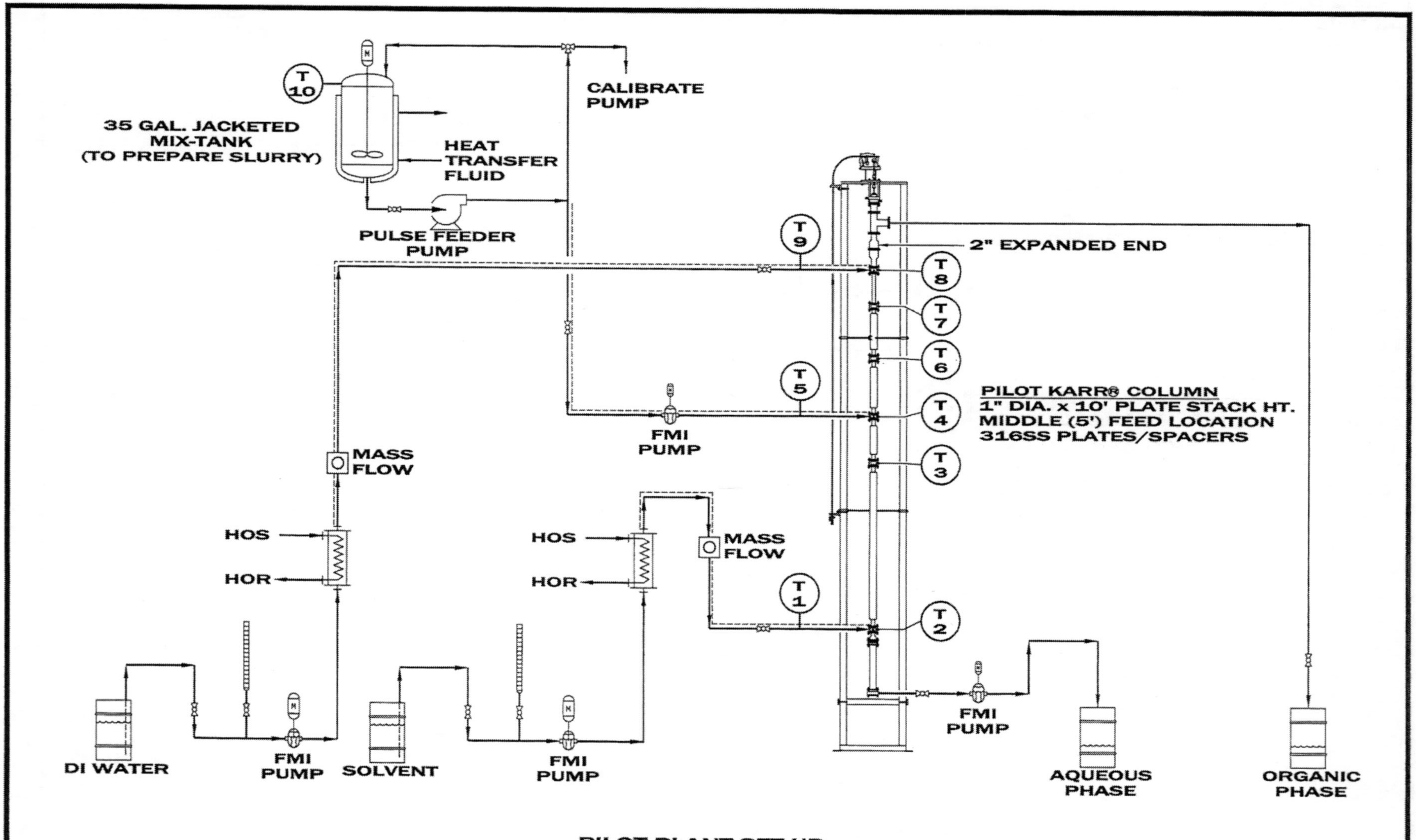

PILOT PLANT SET-UP
FRACTIONAL EXTRACTION WITH MIDDLE SLURRY FEED

FIGURE - 3

KOCH MODULAR PROCESS

551249 The Impact of Low Dynamic Liquid Levels on the Flow Pattern in Pilot Plant Bubble Column Reactor

The Impact of Low Dynamic liquid levels on the Flow Pattern in Pilot Plant Bubble Column reactor

Hayder Al-Naseri[€], J. P. Schlegel*, and Muthanna H. Al-Dahhan[††*♣]

[†]Chemical and Biochemical Engineering Department

[*]Mining and Nuclear Engineering Department

Missouri University of Science and Technology, Rolla, MO 65409, USA

[♣]Cihan University-Erbil, Iraq

[€]Chemical Engineering Department, Tikrit University, Tikrit-Iraq

Abstract

Numerous studies on flow pattern were conducted in lab scale bubble columns with high dynamic liquid level (aspect ratio, $H/D \geq 5$), while in the industry the typical dimension is $H/D \leq 5$. Therefore, the purpose of this work is to study the effect of low aspect ratio ($H/D \leq 5$) on the flow regime transitions in an industrial-sized bubble column. The flow regime at three aspect ratios (H/D = 3, 4, and 5) was demarcated experimentally using linear and non-linear methods, which are represented by the drift-flux and Kolmogorov Entropy (KE), respectively. The four-point optical fiber probe technique has been used to quantify the bubble properties at different regimes and to infer the flow pattern. The experiments were conducted in a bubble column of 0.6 m I.D. and 3.89 m height. The superficial gas velocity varied from 0.005 m/s to 0.45 m/s. The results exhibit that the variation in the aspect ratio has a significant impact on the transition velocity to the churn turbulent regime, while the overall gas holdup in the churn turbulent regime increases with the aspect ratio decreasing. Three mean regimes were indicated by the linear method: bubbly, transition, and churn turbulent. Four regimes are demarcated by the nonlinear method: gas maldistribution, bubbly, transition, and churn turbulent. The results for transition velocity show disagreement with data in the literature. While the empirical correlations of Ribeiro [1], and Şal et al. [2], which are validated with experimental results, introduce a good agreement with percentage errors of 8.6-17.3% and 8.26-25.69%, respectively.

Keywords; Bubble column reactors, Flow regime, Fischer-Tropsch (F-T), Bubble Dynamics

[1] C. P. Ribeiro, "On the estimation of the regime transition point in bubble columns," *Chemical Engineering Journal*, vol. 140, no. 1–3, pp. 473–482, 2008.

[2] S. Şal, Ömer F. Gül, Mustafa Özdemir, Ö. F. Gül, and M. Özdemir, "The effect of sparger geometry on gas holdup and regime transition points in a bubble column equipped with perforated plate spargers," *Chemical Engineering and Processing: Process Intensification*, vol. 70, pp. 259–266, 2013.

551415 Laboratory, Pilot, and Plant Scale Studies in CO_2 Scrubbing

Prior experimentation with CO_2 scrubbing technologies have required caustic alkali hydroxides or toxic amines. To avoid associated corrosion issues and toxicity hazards, the effectiveness of sodium carbonate for capturing CO_2 was tested at laboratory scale. Sodium carbonate is a part of the natural mechanisms by which the ocean captures CO_2 from the atmosphere. We detail the laboratory scale experiments used to determine the potential efficacy of sodium carbonate capture solutions. After promising initial results, we scaled this up to a 4" diameter by 4ft tall packed column, with capturing solution flowing downwards and a simulated flue gas bubbling upwards. We have found that sodium carbonate solution in distilled water with frother modifiers is very effective at capturing CO_2 from a simulated flue gas. At pilot scale using these conditions we are able to capture enough CO_2 that we cannot effectively measure the remaining traces, corresponding to over 98% CO_2 capture. Following these results and to understand the difficulties involved in capturing CO_2 from a real flue gas, we partnered with the Michigan Tech steam plant to install a similar scrubber in their facility. We encountered several distinct problems during this scale up including: the limited availability of distilled water, the presence of CO and other impurities in the flue gas, practical issues involving the appropriate pressurization and filtration of the flue gas, and practical issues involving the influence of pressure and momentum on process kinetics inside the column. We also gained significant insight on the effects of flue gas variability on the capture process. From these issues, we have found that several solution impurities can have a significant negative impact on the performance of the sodium carbonate scrubbing solution – limited time was spent on isolating the specific impurities, but both tap water and softened water were found to have significantly worse capture rates in baseline testing. We have found that the frother modifiers may be deactivated by some of these impurities, requiring additional water preparation or careful frother choice to overcome. We have found that overly high gas flowrates result in gas channels forming inside the column, significantly impeding the CO_2 transfer into the aqueous phase. Despite these challenges, the core process has remained effective even at pilot scale. In this document we report the capture efficiencies of this CO_2 scrubbing process at numerous conditions at each scale of the experimentation, including (where applicable) using distilled water, tap water, softened water, varying concentrations of sodium carbonate, varying concentrations of CO_2 in simulated and real flue gases, and in the presence and absence of various frother chemistries.

552685 Fishing for a Solution: A Hands- on Approach to Continuous Improvement

Key to any continuous improvement effort is having the ability to problem solve your way through the obstacles you encounter. This workshop will give hands-on practice in two techniques that teams can use in problem solving. Join us in this 60 min interactive workshop in exercises you can take back to your teams.

552405 On a Unit Level: Managing Operational Excellence

Continuous improvement is something that is managed at many levels, at a company level, site level, and unit level. This presentation will introduce how it is managed on a unit level or at a lower level where people not only have to buy off on using continuous improvement, but have to use it on a daily basis.

551030 Continuous Precipitation and Agglomeration of Lignin from Organosolv Pulping Liquors at Pilot Scale

Lignin is the largest natural and renewable resource of aromatics on earth. It is one of the three main constituents of lignocellulosic (LC) biomass. However, lignin is still underexploited in the worldwide pulping industry, where lignin is mainly combusted to release its heating value and only a small part is utilized as chemicals and materials [1]. One reason for the current situation is that the conventional pulping processes (e.g. Kraft pulping) were originally not designed to recover lignin as product, although processes for lignin recovery in those mills have been developed [2-5].

Alternative organosolv (organic solvent + water) pulping processes that are designed to recover all the LC biomass constituents as products (pulp, lignin and hemicellulose) are well known for decades [6]. However, commercial organosolv pulping mills, realized in the 1980s and 90s, were shut down again because of problems in downstream processing (solvent cycle closure and lignin recovery) and other issues [7].

In this contribution, a novel continuous process for combined lignin separation and solvent recycling from organosolv pulping liquors is presented [8, 9]. The "LigniSep" process was developed in lab scale at the MPI DCTS and upscaled to a pilot plant at the Fraunhofer CBP in Leuna, Germany. The process is based on the phase behavior of lignin in the solvent/water system (solubility and softening) investigated before [10]. Lignin yield is increased in comparison to dilution precipitation processes by simultaneous precipitation at low solvent concentrations and solvent evaporation. Controlled lignin softening systematically prevents lignin incrustations and triggers lignin particle agglomeration by carefully arranged process conditions. The lignin agglomerates are much better filterable than the precipitated primary particles. The process is monitored and controlled by inline analytical techniques, like infrared spectrometry for solvent content measurement and particle size measurement and microscopy to follow lignin particle formation. A quantitative dynamic model of the pilot plant process was formulated to predict process conditions for given start and boundary conditions. The model was used for the planning of save experiments in the pilot plant. Pulping liquors from batches with different biomass and pulping conditions have been processed in 600kg scale, each. The pulping was performed in another pilot plant right next to the lignin and solvent recovery plant. Results from several experimental campaigns (e.g. process trends, mass balances and yields) will be presented.

1. Cazacu, G., M. Capraru, and V.I. Popa, *Advances Concerning Lignin Utilization in New Materials*, in *Advanced Structured Materials*, S. Thomas, Editor. 2013, Springer: Berlin-Heidelberg. p. 255-312.
2. Zhu, W.Z. and H. Theliander, *Precipitation of Lignin from Softwood Black Liquor: An Investigation of the Equilibrium and Molecular Properties of Lignin.* Bioresources, 2015. **10**(1): p. 1696-1714.
3. Loutfi, H., B. Blackwell, and V. Uloth, *Lignin Recovery from Kraft Black Liquor - Preliminary Process Design.* Tappi Journal, 1991. **74**(1): p. 203-210.
4. Öhman, F., *Precipitation and separation of lignin from kraft black liquor.* 2006, Doctoral thesis Thesis, Chalmers University of Technology.
5. Niemi, H., et al., *Fractionation of Organic and Inorganic Compounds from Black Liquor by Combining Membrane Separation and Crystallization.* Chemical Engineering & Technology, 2011. **34**(4): p. 593-598.
6. Tayenthal, K. and T. Kleinert, *Process of decomposing vegetable fibrous matter for the purpose of the simultaneous recovery both of the cellulose and of the incrusting ingredients.* 1932. US1856567 (A)
7. Pye, E.K. and J.H. Lora, *The Alcell Process - a Proven Alternative to Kraft Pulping.* Tappi Journal, 1991. **74**(3): p. 113-118.
8. Schulze, P., et al., *Method for precipitating lignin from organosolv pulping liquors*. 2014. WO2016062676A1
9. Schulze, P., et al., *Advanced process for precipitation of lignin from ethanol organosolv spent liquors.* Bioresour Technol, 2016. **199**: p. 128-34.
10. Schulze, P., *Lignin Separation from Ethanol Water Pulping Liquors.* 2018, Disseration Thesis, Otto-von-Guericke Universität. 10.13140/RG.2.2.24072.26887

Use of Extractive Separation Technology for Removal of Acid Contaminants from LPG to Meet Product Specifications

Yanxiao Yuan, PhD, Robert D. Barnum, PE, Eric N. Rodano, Jonathan N. Gonzales, and David M. Thomas

Pentair Filtration Solutions, LLC, Conroe, TX

Abstract

Liquefied petroleum gas (LPG) is a typical hydrocarbon product produced by refineries and gas plants with a wide range of commercial and industrial uses. LPG streams are usually sour with contaminants that can lead to pipeline corrosion and cause the product to be off specification. Contaminants commonly found in LPG streams are hydrogen sulfide (H_2S), carbon dioxide (CO_2), and other sulfur compounds such as mercaptans, carbonyl sulfide (COS), carbon disulfide (CS_2) and elemental sulfur. The removal of contaminant species from LPG and other fluids is required in order to significantly reduce corrosion potential from LPG products and make the product suitable for transportation or use in downstream processes.

This study investigated the ability of Pentair's extractive separation technology to remove acid contaminants from an LPG stream that routinely failed the copper strip test. The goal of this study was to determine the root cause of the copper strip test failures and understand the impact of factors such as contact stages and variable contact time on the removal efficiency of each individual contaminant species.

1. Introduction

Liquefied petroleum gas (LPG) is a liquid hydrocarbon stream that is produced in gas plants and refineries with a carbon fraction ranging from C_2 to C_6. The primary hydrocarbon fractions typically found in LPG are propane (C_3) and butane (C_4) [1,2]. LPG has a wide variety of uses, both commercially as a liquid fuel and industrially as feedstock to downstream units, such as an alkylation unit or as feedstock for the petrochemical industry [1,3].

LPG streams are typically sour and contain dissolved acid gases and other contaminants that can lead to pipeline and distribution system corrosion [2,4]. Contaminants commonly found in LPG streams are hydrogen sulfide (H_2S), carbon dioxide (CO_2), and other sulfur compounds such as mercaptans, carbonyl sulfide (COS), carbon disulfide (CS_2) and elemental sulfur [1,5]. H_2S is not only a corrosive substance, it can also lead to the deposition of elemental sulfur in distribution systems [5] and is toxic to humans when released to the atmosphere. While COS and CS_2 alone are not directly responsible for corrosion, both species can hydrolyze in the presence of water to form H_2S and CO_2 [6]. Similarly, mercaptans may not be directly responsible for corrosion [7,8,9,10], but high mercaptan concentrations can cause the LPG stream to be outside the product specifications as feedstock for other processes. Mercaptans are also sources of foul smell in both the LPG product and LPG combustion gases [5]. The amount of each contaminant mentioned above depends largely on the feed source and the type of operation or process upstream of LPG product.

Sour LPG streams must be treated to remove the above mentioned contaminants down to the lowest practical level in order to meet product fuel specifications or to serve as feedstock to downstream processing units. The Gas Processors Association (GPA) – as cited by the Gas Processors Suppliers Association (GPSA) – provides specifications for commercial grade propane, butane and their mixtures (LPG). For contaminant species concentration, four specifications are included – corrosion (copper strip test), total sulfur content, moisture content, and free water content [11,8]. These product specifications are reproduced in Table 1.1 below. Note that, with respect to corrosion, LPG samples must pass the copper strip test with a result of No.1. The product specification for LPG will vary depending on the intended use. For example, most refiners or gas plant operators that treat LPG for export are typically required to meet the following specifications: $H_2S \leq$ 1ppmw, mercaptans $\leq$ 5-10 ppmw, COS $\leq$ 1-2 ppmw, total sulfur $\leq$ 30 ppmw, water $\leq$ 10 ppmw. However, when LPG is used as feedstock for petrochemicals production, the mercaptan specification is $\leq$ 1 ppmw [1].

Table 1.1 - GPA LPG Specifications [11,8]

Product Characteristics	Product Designation		
	Commercial Propane	Commercial Butane	Commercial B-P Mixtures
Corrosion, copper strip, max	No. 1	No. 1	No. 1
Total Sulfur, ppmw	185	185	140
Moisture Content	pass	–	–
Free water content	–	None	None

There are several commercially available technologies capable of removing the previously mentioned contaminants from LPG. Some examples include amine treatment, caustic treatment, and adsorption by molecular sieves or sorbents. Of these methods, treatment of LPG with amines is quite popular within this industry [5,3]. One clear advantage of using amines is relatively low capital cost when removing high levels of H_2S and CO_2 as amines can be regenerated. A disadvantage is that additional separation equipment, such as caustic treating or molecular sieves, may be required to further remove H_2S and other sulfur compounds below product specification levels (including copper strip test results). Additionally, amine systems are not effective solutions for mercaptan removal [1].

Caustic treatment, which is the method employed in this investigation, is very effective in removing most of the various process contaminants, especially mercaptans, as sodium hydroxide (NaOH) is a stronger alkali than amine species. The reaction equations between H_2S, CO_2, mercaptans, and caustic are given below:

$$H_2S + 2NaOH \rightarrow Na_2S + 2H_2O \quad \text{Eq (1.1)}$$

$$CO_2 + 2NaOH \rightarrow Na_2CO_3 + H2O \quad \text{Eq (1.2)}$$

$$RSH + NaOH \rightarrow RSNa + H_2O \quad \text{Eq (1.3)}$$

Caustic treating is typically the preferred choice when mercaptan specifications must be met. However it is not effective for the removal of COS. For processes where COS removal is required, amines are usually added to caustic to help improve the removal of COS [1]. One disadvantage associated with typical caustic treatment systems is the higher cost, particularly when treating LPG with higher contaminant concentrations [12]. The reactions between caustic and acid contaminants – shown above – form stable salts such as Na_2CO_3, NaHS and Na_2S. This makes the caustic solvent difficult to regenerate and can lead to high disposal costs if large caustic rates are required. Caustic treating is most often used to polish trace amounts of acid contaminants from LPG streams.

Molecular sieve technology has a high degree of selectivity for polar compounds such as water, mercaptans, H_2S and CO_2, but a lower selectivity for COS [1]. Molecular sieves require a large capital expenditure and have high operating cost associated with regeneration. Similar to caustic treating, molecular sieves are a viable option to polish trace concentrations of sulfur contaminants from LPG streams.

Each of the technologies described above carry various advantages and disadvantages. Choosing the most appropriate LPG sweetening technology will depend on many factors including product stream contaminant concentrations, product specification requirements, synergy with the gas plant or refinery operations, and environmental considerations.

2. Background

In this study, the chosen test facility is a 20,000 to 25,000 BBL per day topping refinery. The plant produces products such as LPG, gasoline, diesel, kerosene, and atmospheric bottoms/residual gas. The facility currently produces LPG at a rate of 10 to 20 gpm (roughly 500 BBL/day), but as the plant has processed more light, domestic shale crudes, they have seen the LPG product range increase up to 1,500 BBL per day. The operating conditions of the LPG stream are ambient temperature, 120 to 130 psig, CO_2 concentrations as high as 3900 ppmw, H_2S concentrations as high as 300 ppmw, and mercaptan concentrations as high as 150 ppmw.

In order to sell their LPG product on specification, the facility is required to meet the following parameters: 10 ppmw H_2S, 500 ppmw CO_2, no free water, and needs to pass the copper strip test. If the LPG does not meet these parameters, then the LPG is sold as an off specification product at a penalty of \$0.08/gallon. With recent spot prices of LPG ranging from \$1.05 to \$0.62 per gallon (an average of \$0.83 per gallon) [13] the \$0.08 per gallon loss in off specification product is roughly 10% loss in value. This would equate to a loss in profitability of \$1,680 to \$5,040 per day based on the LPG product flowrates seen at the plant.

Due to the presence of acid gases and other contaminants, the plant reported that the LPG stream routinely failed to meet the copper strip test specification. As a result, the refinery was in search of an LPG treatment solution that could reliably ensure adherence to the LPG product specifications, including the copper strip test. Pentair's patented Polarex® technology is an extractive separation system designed to remove entrained polar compounds from hydrocarbon streams with a high degree of extraction efficiency [14,15,16,17]. This technology can be used to replace conventional water wash systems and solvent scrubbing towers [14]. Compared with conventional solutions, Pentair's Polarex® technology provides significantly improved separation performance with less carryover issues, significant reduction in capital and operational costs, significant reduction in footprint and a system that is easier to operate [14]. This technology has already been successfully installed in a variety of hydrocarbon processing applications.

This study investigated the ability of the Polarex® technology to remove acid contaminants from the refinery's LPG product. The ultimate goal was to purify the LPG sample such that it will pass the copper strip corrosion specification (ASTM D1838). This investigation also reviewed how multiple contact stages and variable contact time impacted the removal efficiency of various contaminant species.

3. Methods

A multi-stage pilot testing system was assembled for this study and was connected to the LPG system at the refinery. The pilot configuration consisted of two stages of caustic wash followed by one stage of water wash. The installation of water wash systems downstream of solvent extraction systems is a method commonly used in industry to prevent solvent carryover into product streams [3]. The pilot system was set up in a manner which allowed for the variation of the solvent-product contact time by way of varying the contact volume of each stage. A process flow diagram of the pilot system is provided below in Figure 3.1.

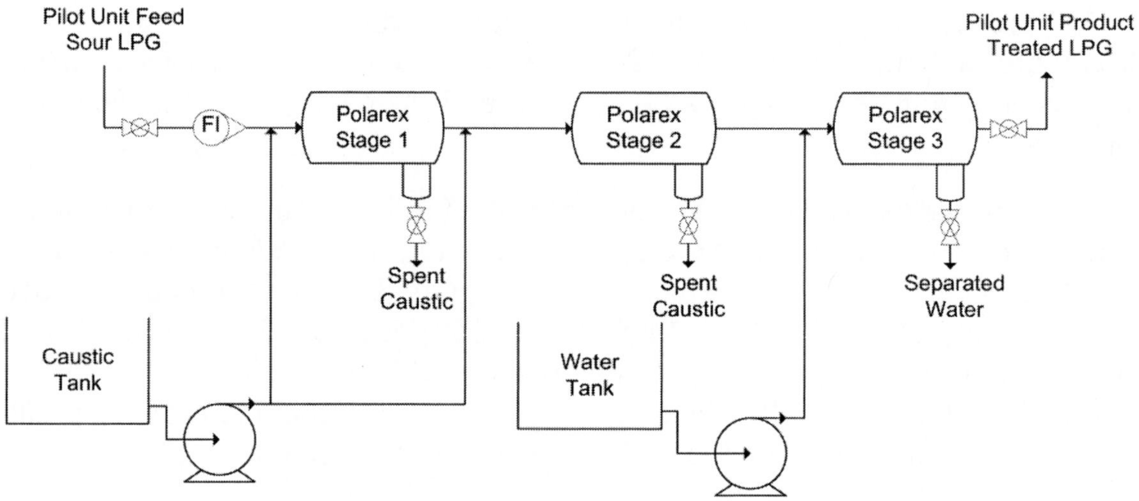

Figure 3.1. Polarex® Pilot Unit Process Flow Diagram

The pilot unit illustrated above was assembled and installed near the LPG production area at the refinery. A slip stream of LPG was fed to the pilot unit at a rate 0.5 gpm. Two chemical metering pumps were used for solvent injection: one for caustic wash for the first two stages and another for water injection into the third stage. Figure 3.2 provides a picture of the pilot unit installation at the refinery.

Figure 3.2. Polarex® Pilot Unit Installation

The pilot testing consisted of three separate test configurations. Each test evaluated a different feed-solvent contact time. Oxidant, amine, and other promoter chemicals were added to the caustic solvent in one test to evaluate the impact on removal of COS. Table 3.1 lists the variables for each test configuration at each stage of the pilot system. The caustic solvent (aqueous NaOH solution) was fed to each stage of the pilot system at an injection rate of 3% by volume of the LPG flow for all the tests.

Table 3.1. Polarex® System Test Configuration and Solvent Selection

Test #	Stage 1		Stage 2		Stage 3	
	Contact Time	Solvent	Contact Time	Solvent	Contact Time	Solvent
Test 1	Long	Caustic	Long	Caustic	Long	Water
Test 2	Short	Caustic	Short	Caustic	Short	Water
Test 3	Long	Caustic w/ oxidant	Long	Caustic w/ oxidant	Long	Water

For each test configuration, the pilot unit was allowed to run until the system reached a steady state. LPG samples were then collected at the pilot system inlet and at the outlet of each contact stage. A copper strip test, and compositional analysis of CO_2, H_2S, mercaptans, elemental sulfur, COS, CS_2, total sulfur and water in the LPG were performed for each sample. The analytical methods used for sample analysis in this investigation are listed in Table 3.2 below.

Table 3.2. Analytical Methods

Analysis	Analytical Method
Copper Strip Test	ASTM D1838
CO_2	GPA-2177
H_2S	ASTM D5623M
Mercaptans	ASTM D5623M
COS	ASTM D5623M
CS_2	ASTM D5623M
Elemental Sulfur	UOP 377
Total Sulfur	ASTM D6667
Water	ASTM E-203

4. Results and Discussion

The pilot system was equipped with a sight glass in the middle of each stage. The primary purpose of the sight glass was to ensure proper separation of the polar and non-polar phases. The sight glasses were also used to visually assess clarity of both the continuous phase (LPG) and the discontinuous phase (spent caustic). Visual clarity assessments can serve as qualitative indicators of contaminant separation efficiency.

Figure 4.1 provides images of the spent solvent and treated LPG in the sight glass for Stage 1, Stage 2, and Stage 3 of the pilot system. It is obvious from the figure that the spent caustic from Stage 1 was darker in color than the caustic from Stage 2, and the washed water was the cleanest. The visual assessment implies that the caustic successfully extracted contaminants, that Stage 1 extracted more contaminants than Stage 2, and that the LPG stream became progressively cleaner after each stage of caustic injection and the water wash. The reduction in each of the acid contaminant concentrations after the pilot system treatment, as will be discussed later, confirms the results of the visual observation of the sight glasses.

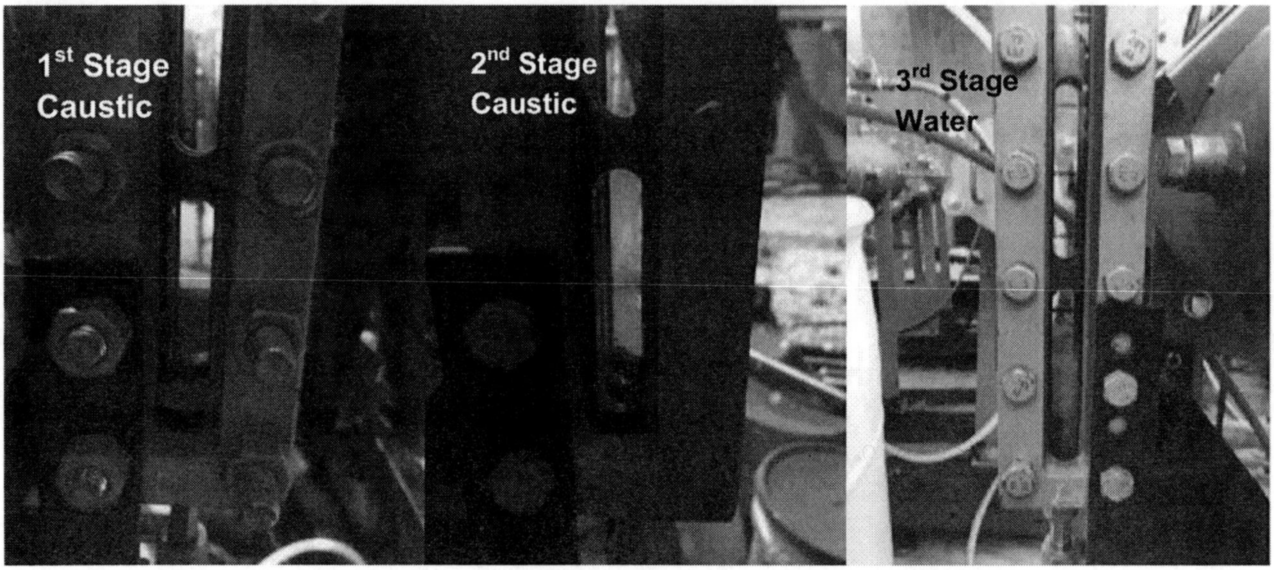

Figure 4.1. Spent Caustic and Wash Water in Sight Glasses

4.1 Copper Strip Test

In order to gauge the effectiveness of any contaminant removal system, industry standard tests must be performed on the processing fluid at various points in the removal process. In the case of corrosion concerns, the primary testing method for corrosivity of LPG is ASTM D1838 [18]. In this test, a polished copper strip is immersed in the product sample for one hour at 100°F. The test strip is then qualitatively rated by comparison with the four color classifications of ASTM D1838. The specification for LPG products stipulates that the product must not receive a copper strip test rating higher than No.1. This rating indicates that the product is free of corrosion causing contaminants.

As mentioned, the LPG product at the test facility routinely failed the copper strip test specification. In order to assess the performance of Pentair's Polarex® technology, the copper strip test was performed on samples from the inlet and outlet of each stage. Performing the copper strip test on the LPG feed to the pilot unit allowed for verification that the LPG stream failed to meet the required specification. Figure 4.2 presents the pilot system copper strip test results. The copper strip test result from the LPG feed sample for each of the three tests was No. 4. This is a strong indicator that corrosive compounds were present in the LPG feed stream.

As shown in Figure 4.2, the copper strip test results from the system outlet sample for each test indicate that Polarex® technology effectively removed corrosive compounds from the LPG stream and produced consistent copper strip test values of No. 1 for the treated LPG product.

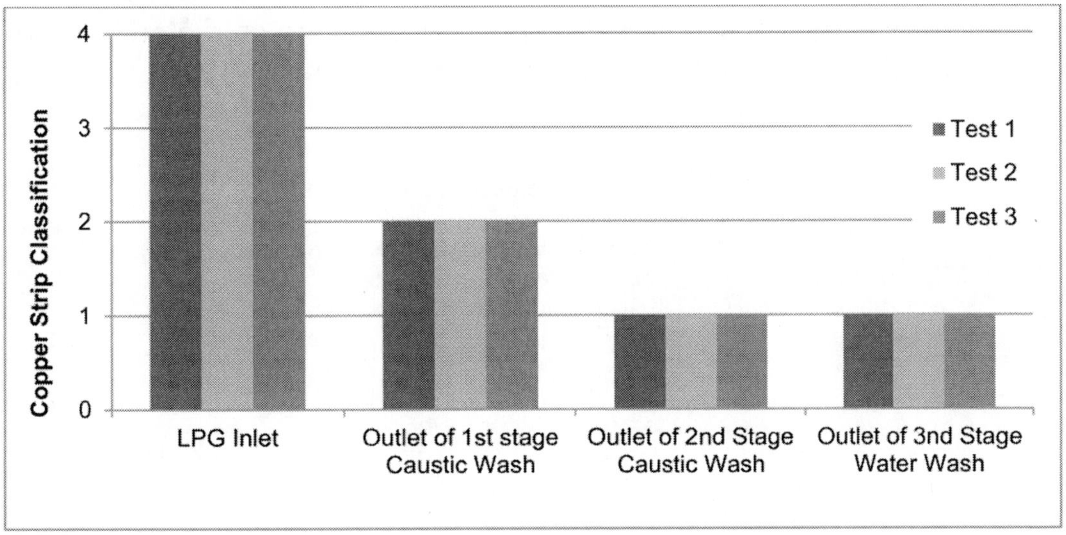

Figure 4.2. Copper Strip Corrosion Test Results

It is also clear from this figure that a multistage caustic wash system was required for complete contaminant removal. Although the first stage of caustic wash was able to reduce the copper strip value from No. 4 to No. 2, the LPG product was still off specification at that point in the system. However after a second stage of caustic injection, the value was reduced to No. 1. The fact that each of the three tests showed the same trend indicates that two stages of caustic injection and extraction were necessary in order to allow the LPG product to meet the copper strip test specification.

The water wash stage was designed to remove caustic residue carried over from previous caustic injections. The data from Figure 4.2 confirms that the water wash stage did not affect the copper strip test results of the effluent stream.

4.2 Hydrogen Sulfide (H_2S)

The copper strip test specification does not state a specific reason for a positive test result, but the industry generally agrees that the likely cause of an off specification test result is caused by corrosion from sulfur compounds present in the hydrocarbon [19]. As discussed, various studies have confirmed that H_2S, elemental sulfur, COS and CS_2 in the presence of water, and organic polysulfides can lead to positive results in copper strip test conducted according to ASTM D1838 [1,5,6,7,8]. These compounds must be removed from LPG down to extremely low levels in order to pass the copper strip test and to meet other product specifications.

In order to determine the root cause of copper strip test failure by the LPG stream at the test facility, it was necessary to quantify corrosive contaminant levels at each sample point in the pilot system. As mentioned previously, H_2S is a highly corrosive compound and is considered to be a primary source of copper corrosion found in LPG streams [1,5,7,8]. Recall that H_2S concentration in the LPG feed stream was expected to be ~300 ppmw. The pilot test results presented in Table 4.1 indicate that the LPG inlet stream H_2S concentration was not as high as initially expected. The highest observed H_2S concentration in the untreated LPG feed was less than 5 ppmw. For Test 2, the feed stream contained less than 0.2 ppmw, which is less than the detection limit of the testing method.

Table 4.1. H_2S in Untreated and Treated LPG

Test #	Analysis	LPG Feed	Outlet of 1st Stage Caustic Wash	Outlet of 2nd Stage Caustic Wash	Outlet of 3nd Stage Water Wash
Test 1	H_2S, ppmw	0.2	<0.2	<0.2	<0.2
Test 2	H_2S, ppmw	<0.2	<0.2	<0.2	<0.2
Test 3	H_2S, ppmw	4.3	<0.2	<0.2	<0.2

As previously mentioned, the LPG feed stream failed the copper strip test, despite the low H_2S concentration in the LPG feed stream. The results of this study do not indicate a direct correlation between H_2S concentration and copper corrosion test results for this particular LPG product stream. However, research by Pyburn et al. (as cited in GPSA) indicate that hydrogen sulfide concentrations of 0.55 ppmw or more in propane almost certainly will produce a failure of the copper strip test while concentrations as little as 0.3 ppmw might fail, depending on interpretation of the test [7,8].

The pilot test results provided in Table 4.1 indicate that concentrations below 0.2 ppmw may still cause a copper strip test failure. While Pyburn, et al. observed failures at slightly higher H_2S concentrations [7,8], the concentrations observed in this study may still have contributed to the overall corrosion results. It is also possible that, in addition to H_2S, there are other compounds or mechanisms that contributed to or were the primary cause of the copper strip test failure. This may be especially true for Test 1 and Test 2, as these results show low or no detectable H_2S while the copper strip test value was as high as No. 4.

Nevertheless, the reaction rate between H_2S and the caustic appeared to be fast. After the 1st stage of caustic injection, all H_2S concentrations were reduced to non-detectable level as shown above in Table 4.1, regardless of the contact time. Therefore, a second stage of caustic wash and longer contact time were not required for the complete removal of H_2S from the LPG feed stream.

4.3 Carbonyl Sulfide (COS)

As discussed, COS in dry LPG does not directly contribute to corrosion in the copper strip test. However, COS can hydrolyze in the presence water and result in the formation of H_2S, which is a source of corrosion [6]. The hydrolysis reaction between COS and water is provided in equation 4.1.

$$COS + H_2O \leftrightarrow H_2S + CO_2 \qquad \text{Eq (4.1)}$$

Figure 4.3 shows that the COS concentration in the untreated LPG feed at the test facility was consistently less than 1 ppmw. The pilot test results indicate that the caustic treatment was only a partially effective method for COS removal. The concentration was slightly reduced to 0.3 ppmw. This observation is consistent with the findings of other studies [1]. Additionally, adding oxidant and other promoters such as amine in Test 3 did not improve the COS removal. Similarly, the water wash step did not appear to cause significant COS hydrolysis as the concentration remained constant after water wash. This is likely due to the low COS concentration and shorter-than-required contact time compared with reported residence times required for the slow COS hydrolysis reaction to progress [6,19].

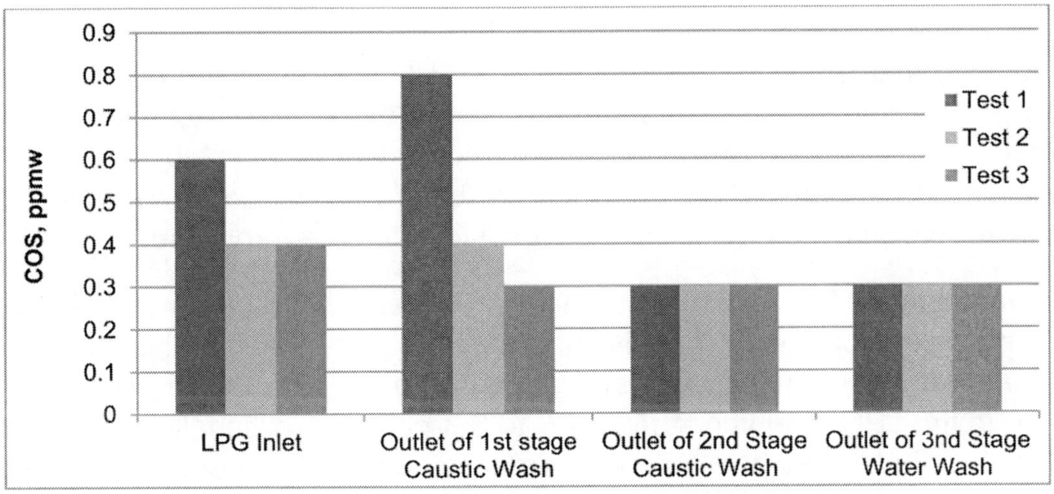

Figure 4.3. COS in Untreated and Treated LPG Streams

4.4 Carbonyl Disulfide (CS_2)

The trend observed from the CS_2 concentration data is similar to the trend for COS concentration. The LPG feed stream contained CS_2 but the concentration was equal to or less than 2 ppmw as shown in Figure 4.4. The caustic wash was not able to effectively remove CS_2 from the LPG. The CS_2 concentration in the LPG stream remained nearly constant across each stage of the pilot test system. Additionally, CS_2 hydrolysis in water was not an apparent contributor to copper strip corrosion. This is likely due to the low level of CS_2 present in the LPG feed. The hydrolysis reaction between CS_2 and water is provided in equation 4.2.

$$CS_2 + 2H_2O \leftrightarrow 2H_2S + CO_2 \qquad \text{Eq (4.2)}$$

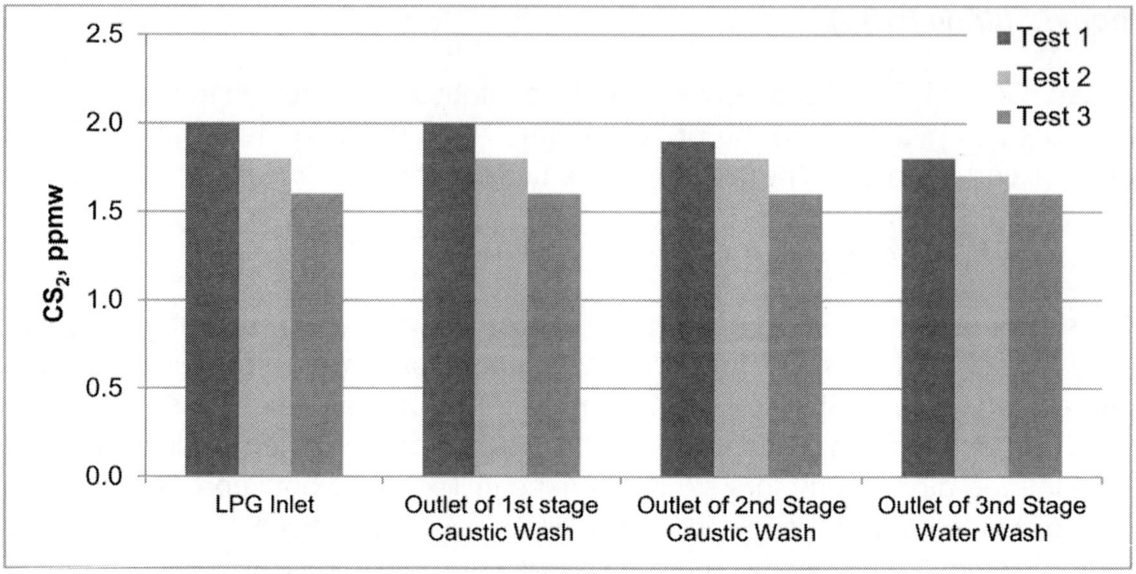

Figure 4.4. CS_2 in Untreated and Treated LPG Streams

4.5 Mercaptans

In this study, total mercaptan concentration in the LPG was analyzed before and after each stage of the pilot system to assess the performance of the Polarex® technology in mercaptan removal. Although mercaptans may not be a direct cause of poor copper strip test results [7,8,9,10], their concentrations do need to be controlled in order to meet the product mercaptan specification if the LPG is used as feedstock to other processes.

The primary mercaptans found in the untreated LPG stream were methyl, ethyl, isopropyl, tert-butyl and n-propyl mercaptans. The total mercaptan concentration (the sum of mercaptan species C4 and smaller) for each test are plotted below in Figure 4.5. The total mercaptan concentration in the LPG feed varied from 4.6 to 9.3 ppmw. Additionally, the results shown in Figure 4.5 confirm that the Polarex® system pilot tests successfully removed mercaptans down to less than 1 ppmw. This level of concentration meets standard product specification for the use of LPG as a feedstock for petrochemical processes. Furthermore, the results from Test 1 and Test 2 indicate that, for the LPG feed conditions and composition present in the refinery, one stage of caustic wash was sufficient to remove mercaptans below the product mercaptan specification.

It is interesting to note that contact time had a substantial impact on mercaptan removal efficiency. The longer contact time in Test 1 resulted in a reduction in total mercaptan concentration from 9.3 ppmw at the inlet to below the detection limit after the 1st stage of caustic wash. The removal efficiency for this stage was nearly 100%. By comparison, Test 2 experienced a reduction in total mercaptan concentration from 4.6 ppmw to 0.9 ppmw across the 1st stage, a removal rate of 80%. The system configuration of Test 2 was identical to Test 1, with the exception of shorter contact time. This implies that the reaction between mercaptans and caustic is slower compared to the reaction between H_2S and caustic, which is expected. These results further indicate that having a second stage of caustic wash may not be required to achieve the desired mercaptan removal if the contact time is sufficiently long. However, if the total mercaptan concentration in the LPG feed is higher than the concentration in this study, a secondary stage of caustic wash may still be required to meet the product specification of less than 1 ppmw.

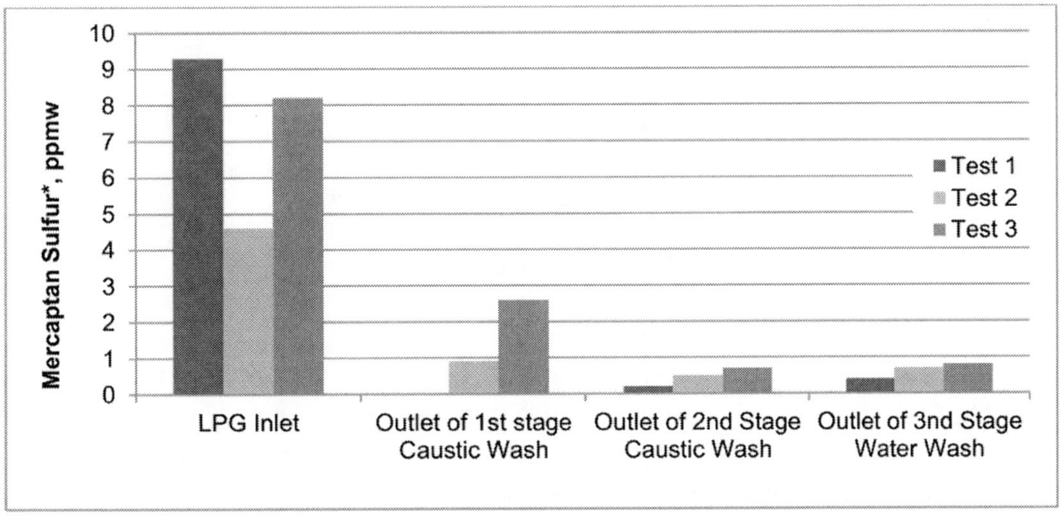

Figure 4.5. Mercaptans in Untreated and Treated LPG Streams

The 1st stage of caustic wash successfully removed the majority of the mercaptans from the LPG. However, the LPG samples still failed the copper strip test. These results indicate that mercaptans were not a cause of corrosion after the 1st contact stage. This is especially true for Test 1, where mercaptans were removed below the detection limit. Based on the pilot test results, it is likely that mercaptans are not a cause of corrosion in this study, which is consistent with the findings from others [7,8,9,10].

4.6 Elemental Sulfur

In addition to H_2S, elemental sulfur is a sulfur contaminant that can cause off specification copper strip test results. It has been reported that the appearance of elemental sulfur in LPG streams is less common than other contaminants [1]. However, elemental sulfur was detected in the untreated LPG feed stream samples from the test facility. The elemental sulfur concentration in the untreated LPG feed samples varied from 4 to 60 ppmw during the period of pilot testing as illustrated below in Figure 4.6.

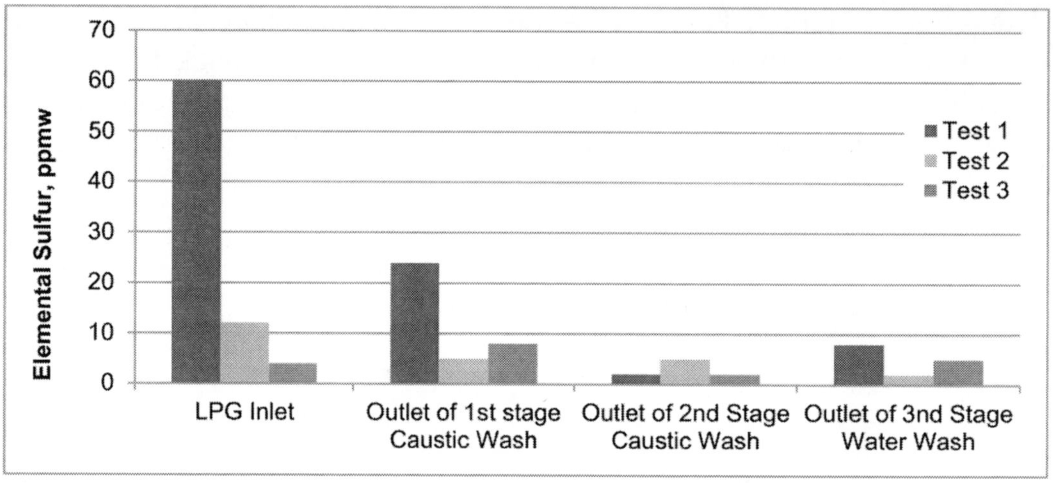

Figure 4.6. elemental sulfur in Untreated and Treaded LPG

The results of this study indicate that the Polarex® system effectively removed elemental sulfur from the LPG product stream. A reduction in the elemental sulfur concentration was seen after the 1st stage for both Test 1 and Test 2. No clear reduction was seen for Test 3, likely due to the error in measurement of low elemental sulfur concentrations combined with low accuracy of the test method for elemental sulfur. The overall elemental sulfur removal efficiency for the 3-stage Polarex® system was 87% for Test 1 and 83% for Test 2. Figure 4.6 shows that the elemental sulfur concentration in the effluent LPG stream was reduced to a 2 to 5 ppmw level. For all three tests, the effluent LPG samples were corrosion free with a copper strip test rating less than No. 1.

While H_2S and mercaptans were removed by a single stage of caustic wash, the same is not true for the removal of elemental sulfur. After the first stage of caustic wash, elemental sulfur was reduced to 5 to 24 ppmw. The corresponding copper strip test results were No. 2 for each test. Research by Pyburn et.al suggests that the corrosive threshold for elemental sulfur by itself is approximately 2 to 10 ppmw [7,8]. Recall also that at this point in the pilot test system, the H_2S concentration was reduced below the analytical detection limit in every test. Mercaptans were also completely removed by the first stage when a long reaction time was provided. Hydrolysis of COS and CS_2 did not contribute to the corrosion. Combining all of this information, the results indicate that elemental sulfur alone caused copper corrosion when the concentration was above a certain level. This conclusion is consistent with the findings of Pyburn, et. al.

Research also suggests that elemental sulfur in the presence of H_2S results in a synergistic effect that will cause failure of the copper strip test with trace concentrations of either contaminant [7,8]. Specifically, concentrations of 0.5 ppmw elemental sulfur, together with 0.28 ppmw H_2S in propane were observed to fail the corrosion test. The threshold value of hydrogen sulfide may be as low as 0.12 ppmw in the presence of 0.5 ppmw elemental sulfur. With respect to this investigation, however, synergistic effects cannot be confirmed to be a significant cause of corrosion. While the synergy may have had a secondary corrosion affect in the untreated LPG feed, H_2S was removed to below the detection limit after the first stage of caustic wash. When the elemental sulfur concentration was reduced to below the corrosive threshold after the second stage, copper corrosion no longer occurred. These results confirm that significant H_2S concentrations were no longer present and synergy did not occur between the contaminants. This again suggests that elemental sulfur was most likely the main source of copper corrosion for this investigation.

4.7 Total Sulfur

In addition to the concentrations of various individual sulfur compounds, total sulfur concentration was also measured across the pilot test unit. This data could provide some additional insight into the overall sulfur-based contaminant removal efficiency of the pilot system. Figure 4.7 presents the total sulfur concentrations before and after each stage of the Polarex® system. As before, a consistent reduction in total sulfur was observed across the entire Polarex® system. The results for Test 1 show that the total sulfur concentration was reduced from 15.0 ppmw at the LPG feed to 3.5 ppmw after the 2nd stage of caustic injection. This equates to a removal efficiency of 77%. For Test 2, the concentration was reduced from 9.3 ppmw to 3.4 ppmw after the 2nd stage of caustic injection which is a removal efficiency of 63%. No clear trend was observed for Test 3 since the initial concentration was already lower than 4 ppmw.

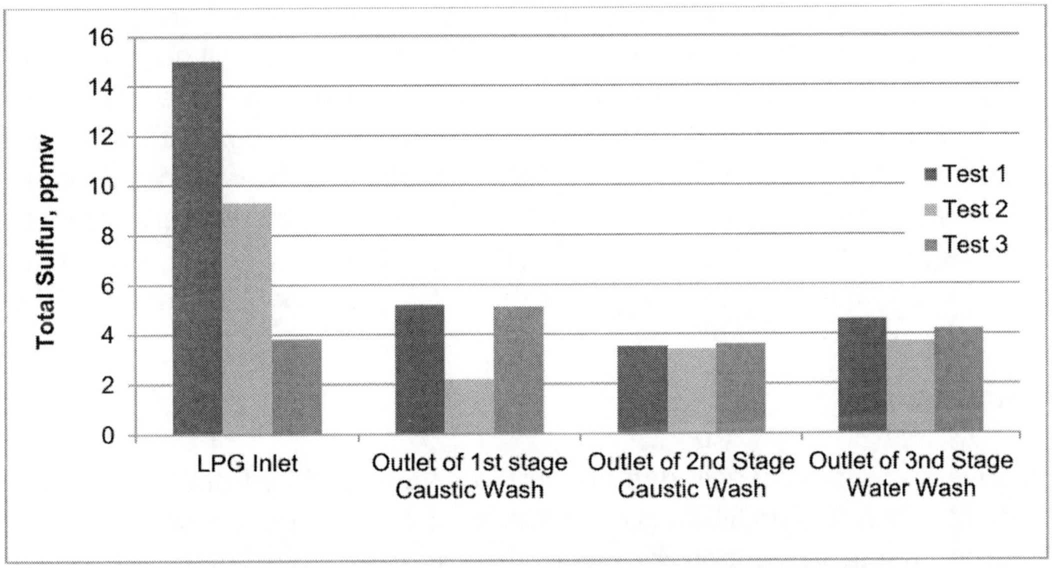

Figure 4.7. Total Sulfur in Untreated and Treaded LPG

In order to provide a more clear comparison of all pilot test results for the various sulfur species considered in this study, Table 4.2 summarizes the removal efficiencies of all the sulfur species analyzed. As shown in the table below, total sulfur removal efficiency was 65% to 76% across Stage 1 and further reduced by another ~30% by Stage 2. The overall system removal efficiency was 60% to 69%.

Table 4.2. Sulfur Species in Untreated and Treated LPG

Test #	Sulfur Species	LPG Feed, ppmw	Outlet of 1st Stage		Outlet of 2nd Stage		Outlet of 3rd Stage		System Efficiency
			Conc., ppmw	Efficiency	Conc., ppmw	Efficiency	Conc., ppmw	Efficiency	
Test 1	H_2S	0.2	<0.2	–	<0.2	–	<0.2	–	–
	Elemental S	60	24	60.0%	2	36.7%	8	-10%	86.7%
	Mercaptans	9.3	0	100.0%	0.2	-2.2%	0.4	-2%	95.7%
	Total S	15	5.2	65.3%	3.5	11.3%	4.6	-7%	69.3%
Test 2	H_2S	<0.2	<0.2	–	<0.2	–	<0.2	–	–
	Elemental S	12	5	58.3%	5	0.0%	2	25%	83.3%
	Mercaptans	4.6	0.9	80.4%	0.5	8.7%	0.7	-4%	84.8%
	Total S	9.3	2.2	76.3%	3.4	-12.9%	3.7	-3%	60.2%
Test 3	H_2S	4.3	<0.2	>95%	<0.2	–	<0.2	–	>95%
	Elemental S	4	8	-100.0%	2	150.0%	5	-75%	-25.0%
	Mercaptans	8.2	2.6	68.3%	0.7	23.2%	0.8	-1%	90.2%
	Total S	3.8	5.1	-34.2%	3.6	39.5%	4.2	-16%	-10.5%

As noted in previous sections, the Polarex® system was able to remove specific sulfur species efficiently. Elemental sulfur was removed by the pilot test system with an efficiency between 83% and 87%. The elemental sulfur concentration was reduced to 2 to 8 ppmw. This concentration level may be the threshold at which elemental sulfur will no longer be a source of corrosion on copper. The results could also indicate that at the observed elemental sulfur concentration levels no synergistic effect existed when the H_2S concentration was below 0.2 ppmw. Additionally, the Polarex® system with caustic wash was able to remove light mercaptans (C4 and smaller) to less than 1 ppmw. Light mercaptans were removed by 68% to 100% across Stage 1, and removed another 44% to 73% by Stage 2. The overall removal efficiency of C4 and lighter mercaptans across the entire Polarex® system was 85% to 96%.

4.8 Carbon Dioxide (CO_2)

Carbon dioxide is also a typical acid gas present in a variety of hydrocarbon streams, including LPG. In situations where a significant concentration of water is not present in the process fluid, CO_2 does not contribute to copper corrosion. The species does, however, react with NaOH and therefore competes with H_2S and other sulfur species for caustic consumption. The table below lists the CO_2 concentration at each point in the Polarex® treatment system.

Table 4.3. CO_2 in Untreated and Treated LPG

Test #	Analysis	LPG Feed	Outlet of 1st stage Caustic Wash	Outlet of 2nd Stage Caustic Wash	Outlet of 3nd Stage Water Wash
Test 1	CO_2, ppmw	1,300	<100	<100	<100
Test 2	CO_2, ppmw	1,300	<100	<100	<100
Test 3	CO_2, ppmw	1,400	300	<100	<100

As was the case for H_2S, the results above indicate that the reaction between CO_2 and NaOH occurred quickly and did not require a long contact time to progress to completion. For a solvent with caustic alone, as was the case in Tests 1 and 2, a single stage was sufficient to reduce the CO_2 concentration from 1,300 ppmw to below detection limit of 100 ppmw. The removal efficiency of a single stage of caustic wash was higher than 92%. Adding other chemicals into the caustic did not improve the performance for Test 3.

4.9 Water

Another interesting finding from the results provided in Table 4.2 is that the water wash step appeared to make the overall system removal efficiency of the sulfur species slightly worse in general. This trend is evident for all the sulfur species analyzed in this study. Hydrolysis of sulfur compounds other than COS and CS_2 may have occurred during the water wash step. The levels of COS and CS_2 were fairly constant after the water wash step. However, the water wash step is necessary in order to remove any entrained caustic from the LPG stream. This is especially true when the required caustic concentration is high.

In this study, the initial water concentration in the LPG was 60 ppmw for all the three inlet samples. After caustic injection and water injection, no additional water was observed to be added to the LPG product. The water concentration remained relatively constant, varying between 50 and 60 ppmw throughout all of the tests. These results show that the Polarex® system separated the injected caustic solvent from the LPG stream after contaminant extraction was complete. In general, Polarex® technology is able to remove liquid in a discontinuous phase to a level close to the soluble concentration. Therefore, the pilot test results indicate that that the water solubility in the LPG stream was slightly below 60 ppmw.

5. Conclusions

The pilot testing results indicate that Pentair's Polarex® technology is a very effective method to chemically extract sulfur compounds from an LPG stream. The results of this investigation show that a multi-stage unit with two stages of caustic wash followed by one stage of water wash was able to remove corrosive compounds to a significant degree, causing the LPG stream to meet the copper strip test specification. This conclusion was reached due to the consistent results of all three tests, each performed with varying test parameters.

In addition to the success in removing corrosive sulfur compounds to pass the copper strip test, the results of this study demonstrate that the Polarex® technology utilizing a caustic solvent wash is able to effectively remove mercaptans from an LPG stream. A stand-alone Polarex® unit can reduce the total mercaptan concentration below 1 ppmw, meeting the product specification as a feedstock for petrochemical processes.

A key focus of this study is the ability of the Polarex® system to remove various corrosion causing contaminants from the LPG stream and to identify the main source of corrosion in the LPG stream. While H_2S is a typical source of copper corrosion in LPG distribution systems, for this investigation the H_2S concentration was found to be fairly low in the untreated LPG feed samples. H_2S levels of 0.2 ppmw or lower were observed in two of the three inlet samples from the pilot tests. Additionally, H_2S was removed to below the detection limit after the first stage of caustic contact in each test run. Therefore, it is likely that other species or mechanisms may have been a significant source of the observed corrosion.

The COS and CS_2 concentrations were found to be relatively low in the analyzed LPG samples. Water hydrolysis of COS and CS_2 did not appear to contribute to the copper strip test.

Elemental sulfur, although typically less common in LPG streams than other sulfur based contaminants, was found to be the main source of the copper strip test failures. Pentair's Polarex® technology used in the pilot setup was able to successfully reduce elemental sulfur concentrations to a degree that the treated LPG product could easily pass the copper strip test specification. Analytical results of this examination show the removal efficiency across the pilot system was 83% to 87%. In each test, the elemental sulfur concentration at the outlet of the 2nd stage was reduced to 2 to 5 ppmw.

The complete elemental sulfur removal at the outlet of the pilot test system correlated to an overall copper strip test rating below No. 1 at the same point. In every test case, two stages of caustic wash were necessary in order to reduce the copper strip test results from No. 4 to No. 1. The pilot test data showed, therefore, that the 2nd stage of caustic wash was necessary for this LPG stream to pass the copper strip test specification.

When considering the impact of contact time on system performance, the results of this study indicate that the reactions between H_2S / CO_2 and the caustic solvent occurred quickly and that increased contact time was not necessary for these reactions to continue to completion. The removal efficiency of elemental sulfur was also not impacted by the contact time. The mercaptan removal efficiency, however, was significantly influenced by the contact time. Increased contact time resulted in a higher mercaptan removal efficiency for that stage. The overall mercaptan removal efficiency across the pilot test system was increased from 85% to 96% when contact time was increased.

6. Copyright and Intellectual Property Notice

7. References

1 Hati AS. Sour LPG Treating. In: Laurence Reid Gas Conditioning Conference proceedings; 2012; Norman, OK. p. 243-260.

2 Jayakumar K, Panda RC, Panday A. A Review: State-of-the-Art LPG Sweetening Process. International Journal of Chemical Engineering Research. 2017;9(2):175-206.

3 Nielsen RB, Rogers J, Bullin JA, Duewall KJ. Treat LPGs with Amine. Hydrocarbon Processing. 1997 Sep:49-59.

4 Smith SN, Joosten MW. Corrosion of Carbon Steel by H2S in CO2 Containing Oilfield Environments. In: CORROSION conference proceedings; 2006; San Diego, CA. Available from: https://www.onepetro.org/conference-paper/NACE-06115.

5 Holmes JW, Spears ML, Bullin JA. Sweetening LPG's with Amines. Chemical Engineering Progress. 1984 May.

6 Perry CR. Treating Systems for Ethane Recovery Plants. In: Gas Conditioning Conference proceedings; 1977; Norman, OK.

7 Pyburn CM, Cahill FP, Lennox RK. The Effect of Sulfur Compound Interactions on the Copper Corrosion Test in Propane. In: Gas Processors Association 57th Annual Convention proceedings; 1978. p. 46-51.

8 Gas Processors Suppliers Association. Product Specifications. In: Engineering Data Book. 12th ed. Vol I. Tulsa (OK): Gas Processors Suppliers Association; 2004.

9 Monzó J, García-Antón J, Guiñón JL. Influence of elemental sulfur and mercaptans on corrosion of copper strips in the ASTM D-130 test by means of electronic microscopy (SEM) and energy dispersive X-ray (EDX). Fresenius' Journal of Analytical Chemistry. 1991;341(10):606-610.

10 Jędrychowska S. Badanie wpływu dodatku różnych związków siarki na działanie korodujące na miedź w paliwie LPG (Study of the effects of the addition of various sulphur compounds on the corrosion of copper in LPG fuel). Nafta-Gaz. 2013;69(9):695-701.

11 Gas Processors Association. 2140-97 Liquifed Petroleum Gas Specifications. Gas Processors Assocaiton; 1997.

12 Fleming KB, Spears ML, Bullin JA. Design Alternatives for Sweetening LPG's and Liquid Hydrocarbons with Amines. In: Society of Petroleum Engineers 63rd Annual Technical Conference proceedings; 1988; Houston, TX.

13 U.S. Energy Information Administration. Mont Belvieu, TX Propane Spot Price FOB. [Internet]. [cited 2019 Feb 15]. Available from: https://www.eia.gov/dnav/pet/hist/EER_EPLLPA_PF4_Y44MB_DPGD.htm.

14 Pentair. Extractive Separations - Polarex. [Internet]. [cited 2018 Jul 10]. Available from: https://oilandgasseparations.pentair.com/en/Technologies/Polarex.

15 Thundyil M, Spearman M, Miller M, Mittlesteadt D, Burns H, Hahn C, inventors. Method for the selective extraction of acids. 2004 Nov 22 (Filed); 2009 Oct 20 (Issued). United States patent US 7,604,685.

16 Thundyil M, Spearman M, Miller M, Mittlesteadt D, Burns H, Hahn C, inventors. Method for the selective extraction of acids, bases and polar salts. 2007 Oct 30 (Filed); 2013 Apr 23 (Issued). United States patent US 8,425,663.

17 Thundyil M, Hahn C, Mittlesteadt D, Burns H, Miller M, inventors. Method for the selective extraction of acids, bases and polar salts. 2007 Feb 16 (Filed); 2013 Sept 10 (Issued). United States patent US 8,529,667.

18 ASTM. D1838−16 Standard Test Method for Copper Strip Corrosion by Liquefied Petroleum (LP) Gases. ASTM; 2016. DOI:10.1520/D1838-16.

19 Clark PD, Lesage KL. The Copper Strip Corrosion Test - Understanding and Supplementing It. In: Laurence Reid Gas Conditioning Conference proceedings; 1991; Norman, OK.

551124 Advanced Non-Invasive Radioactive Particle Tracking (RPT) and Dual-Sources Gamma Ray Computed Tomography (CT) Techniques for 3D Hydrodynamics Measurement for Multiphase Flow Systems (Visualization and Quantification)

Multiphase flow systems are extensively used in chemical and biochemical industries. Where perfect mixing and ideal mass and heat transfer are crucial for high efficient operations. Because the local flow dynamics parameters are the main factor for multiphase systems effective reactor design and successful scale-up. Consequently, these parameters require a suitable understanding through the reactor running stages. Unfortunately, these parameters need to be further developed and evaluated to account of the flow dynamics field in each location inside the reactor which is unclear yet and need more information for deeply understanding. Therefore, in this work, 3D local hydrodynamics have been measured extensively for different multiphase flow systems such as, bubble/slurry bubble (lab and pilot plant scales) with and without heat exchanging tubes, biochemical (cylindrical photobioreactor), spouted bed (different scales), fluidized bed (lab and pilot plant scales), and trickle bed columns by using a non-invasive advanced radioactive particle tracking (RPT) (for lab and pilot scales) and dual-sources gamma-ray computed tomography (CT) techniques. In our laboratory (Multiphase Flow Reactors and Applications Laboratory MFReal), the diagnostics techniques have been well established and used extensively for monitoring the flow field in a non-invasive manner in opaque multiphase reactor systems. The experimental results and conclusions will present at the conference.

550159 Towards the Incorporation of Operability and Safety in the Synthesis of Intensified Reactive and Extractive Separation Systems

Intensified processes typically feature higher degree of integration of multifunctional phenomena, less degrees of freedom, narrower operability windows, and even faster dynamics [1-2]. Thus, the incorporation of control, operability, and safety assessment at the design level becomes even more critical to ensure the actual operational performances in these systems, comparing to their conventional unit-operation-based counterparts. Despite recent advances [3-6], systematic synthesis approaches and analytical tools for the design and operability/safety optimization of intensification processes are still lacking.

To address this challenge, in this work and as part of the RAPID SYNOPSIS Project [7], we propose a systematic framework for the synthesis of operable process intensification systems, with specific focus on reactive and extractive separation processes. This framework is based on a phenomenological process intensification/synthesis approach (i.e., Generalized Modular Representation Framework (GMF) [8-10]), which first identifies candidate/promising intensified tasks and then translates them to equipment-based flowsheet alternatives. Flexibility analysis is integrated with the GMF model to ensure that resulting design configurations can be operable under varying operating conditions. To systematically account for inherent safety performance, risk assessment criteria are included as process constraints considering failure frequency and consequence severity. Simultaneous design and control strategies are also employed to ensure validated operable intensified design configurations using advanced multi-parametric model-based predictive control (mp-MPC) developed via the PAROC (i.e., PARametric Optimization and Control) framework [11]. Two case studies are presented to highlight the potential of the proposed approach: (i) a reactive separation system - methyl-tert-butyl-ether production, and (ii) an extractive separation system - methanol/water azeotropic separation.

References

1. Tian, Y., Demirel, S. E., Hasan, M. F., & Pistikopoulos, E. N. (2018). An Overview of Process Systems Engineering Approaches for Process Intensification: State of the Art. Chemical Engineering and Processing-Process Intensification, 133, 160-210.

2. Nikačević, N. M., Huesman, A. E., Van den Hof, P. M., & Stankiewicz, A. I. (2012). Opportunities and challenges for process control in process intensification. Chemical Engineering and Processing: Process Intensification, 52, 1-15.

3. Baldea, M. (2015). From process integration to process intensification. Computers & Chemical Engineering, 81, 104-114.

4. Carrasco, J. C., & Lima, F. V. (2018). Bilevel and parallel programing based operability approaches for process intensification and modularity. AIChE Journal, 64(8), 3042-3054.

5. Nemet, A., Lee, K., Klemeš, J. J., Varbanov, P. S., Moon, I., & Kravanja, Z. (2015). Safety analysis embedded in total site synthesis. In Chemical Engineering Transactions. Italian Association of Chemical Engineering-AIDIC.

6. Medina-Herrera, N., Grossmann, I. E., Mannan, M. S., & Jiménez-Gutiérrez, A. (2014). An approach for solvent selection in extractive distillation systems including safety considerations. Industrial & Engineering Chemistry Research, 53(30), 12023-12031.

7. SYNOPSIS - Synthesis of operable process intensification systems (DE-EE0007888-09-03).

8. Papalexandri, K. P., & Pistikopoulos, E. N. (1996). Generalized modular representation framework for process synthesis. AIChE Journal, 42(4), 1010-1032.

9. Ismail, S. R., Proios, P., & Pistikopoulos, E. N. (2001). Modular synthesis framework for combined separation/reaction systems. AIChE Journal, 47(3), 629-649.

10. Tian, Y., Mannan, M. S., & Pistikopoulos, E. N. (2018). Towards a systematic framework for the synthesis of operable process intensification systems. In Computer Aided Chemical Engineering (Vol. 44, pp. 2383-2388). Elsevier.

11. Pistikopoulos, E. N., Diangelakis, N. A., Oberdieck, R., Papathanasiou, M. M., Nascu, I., & Sun, M. (2015). PAROC—An integrated framework and software platform for the optimisation and advanced model-based control of process systems. Chemical Engineering Science, 136, 115-138.

550636 Building Block-Based Design and Intensification of Chemical Processes

Process intensification provides means for more sustainable, cost effective, smaller and safer designs compared to their conventional counterparts [1-2]. Often times, intensification is realized through capturing the synergy between different process phenomena. Identification of such intensification opportunities can be facilitated through a departure from unit-operation-based design paradigm which has been the modus operandi since the last century. Accordingly, several process intensification methods have been developed in the past that rely on process phenomena, mass/heat exchange modules and/or tasks [3-4]. However, it remains a challenge to systematically identify the optimal intensified flowsheets for a given design problem. To this end, our recently proposed building block-based approach [5] provides unique advantages for systematic process design and intensification [6-8]. In this work, we will show that this representation approach leads to an optimization-based systematic method for process intensification. Unlike the traditional superstructure-based approaches, our proposed building block superstructure relies on physicochemical phenomena to automatically generate optimal intensified flowsheets. It is formed by building blocks positioned on a two-dimensional grid and connectivity between the blocks is obtained through inter-block mass and energy transfer. By using either single or multiple building blocks, many physicochemical phenomena can be represented. This results in a mixed-integer nonlinear optimization (MINLP)-based model that describes the superstructure. We will demonstrate the benefits of the building block-based design approach in terms of systematic process intensification through several case studies. Special focus will be given to intensification of natural gas utilization systems, energy-intensive separations, and reactive separation processes.

[1] Stankiewicz, A. I., Moulijn, J. A. (2000). Process Intensification: Transforming Chemical Engineering. Chemical Engineering Progress 1, 22–34.

[2] Tian, Y., Demirel, S. E., Hasan, M. M. F., Pistikopoulos, E. N. (2018). An Overview of Process Systems Engineering Approaches for Process Intensification: State of the Art. *Chemical Engineering and Processing: Process Intensification*, 133, 160-210.

[3] Lutze, P., Babi, D.K., Woodley, J.M. and Gani, R. (2013). Phenomena based methodology for process synthesis incorporating process intensification. *Industrial & Engineering Chemistry Research*, *52*(22), pp.7127-7144.

[4] Papalexandri, K. P., Pistikopoulos, E. N. (1996). Generalized modular representation framework for process synthesis. *AIChE Journal*, *42*(4), 1010-1032.

[5] Demirel, S. E., Li, J., and Hasan, M. M. F., (2017). Systematic Process Intensification using Building Blocks, *Computers and Chemical Engineering*, 105, 2-38.

[6] Li, J., Demirel, S.E. and Hasan, M.M.F., 2018. Process synthesis using block superstructure with automated flowsheet generation and optimization. AIChE Journal, 64(8), 3082-3100.

[7] Li J., Demirel S.E., Hasan M.M.F. Process Integration using Block Superstructure. *Industrial & Engineering Chemistry Research*, 2018, 57: 4377–4398.

[8] Li J., Demirel S.E., Hasan M.M.F. Fuel Gas Network Synthesis Using Block Superstructure. *Processes*. 2018, 6(3): 23.

549876 Semi-Continuous Processing to Improve Safety and Quality

Challenges exist for processing viscous products that are classified as flammable due to high solvent content. This paper will focus on the processing options for mixing a highly filled shear thinning base material with a low viscosity flammable solvent followed by direct filling into final packaging.

This process can be accomplished using a change can batch mixer, but with significant drawbacks. Injecting hazardous and/or flammable materials through a hand held cartridge device into a pressurized vessel, such as a change can mixer, has potential for operator exposure and relies on personal protective equipment for protection. Another drawback is that batch processing results in a large quantity of work in progress (WIP) material with potential for failure of an entire batch leading to high scrap costs. To avoid these problems an alternate process was sought out to allow for in process adjustments to hit quality targets while minimizing scrap and improving safety.

Multiple solutions were considered. Use of a continuous mixing unit such as a twin screw extruder or pin mixer was examined. Although the mixing in these units was sufficient and provided higher processing rates, material properties were not easily adjusted. A semi-continuous option utilizing a series of static mixers fed through volume controlled feed units with direct filling into the final packaging was then explored. Several potential hurdles were overcome through multiple trials demonstrating the ability to use this process scheme to achieve our processing goals.

By using the semi-continuous process scheme it was possible to reduce the WIP material by 70% compared to the batch change can mixer. Quality was improved by the ability for in-process testing and process productivity increased. Potential chemical exposure for operators was reduced through an upgraded design of the feed systems.

551424 A Tale of Two Columns: Integrating Dynamic Simulation in Conceptual Design

Steady-state simulators in conjunction with standard column sizing methodologies have been used extensively in industry to design distillation columns. Steady-state simulators, when based on reliable underlying physical property data and models, are invaluable in identifying optimal designs and configurations for distillation. However, steady-state simulations do not consider the impact of hold-ups, operation in certain regimes e.g. vacuum, or commonly encountered perturbations such as feed conditions and utilities. Dynamic simulation has been used widely in industry as a tool to diagnose operability issues for operating plants and equipment and less commonly in earlier stages of process design.Dynamic simulation can also be used as an effective screening tool to identify potential control schemes. Two case studies involving new intensified designs for solvent recovery separations are presented, where dynamic simulation is used as a tool to evaluate operability in the conceptual design stage.

551020 Optimization of Mixed Smr Plant Using Advanced Modelling & Six Sigma Tools to Improve Energy Efficiency and Plant Performance: Case Study

Synthesis gas is the key feedstock for many industrial applications especially for ammonia, methanol, Oxo-alcohols etc. Most of the synthesis gas produced in the world is from natural gas using steam methane reforming (SMR). Mixed SMRs have carbon di-oxide added to the feed gas to produce synthesis gas with lower H2/CO ratio than that of conventional synthesis gas applications. As the gas reserves are depleting in the world, existing reforming plants are open to use natural gas from different sources. Different natural gas sources offer different compositions in the gas, which significantly affect the performance of the mixed steam methane reformers (SMR) in terms of reforming efficiency, energy consumption, H2/CO ratio in synthesis gas etc. Effect of feed gas disturbances in mixed SMRs is significant as these plants are operated at higher reforming temperature and lower steam to carbon ratios. In order to operate the reforming plant efficiently, optimally and smartly for all feed gas disturbances, optimization of the plant using advanced modelling & six sigma tools, precise pro-active control, continuous operations monitoring are necessary.

In this paper, an implemented optimization of mixed steam reformer in the oxo-alcohol production process case study is presented. An Aspen Plus model based study has been performed to optimize the mixed SMR plant operations for the existing feed gas composition. Then, coupled the model-based study with six sigma tools to develop the best operating conditions and zone for various scenarios. A number of contour diagrams have been developed for these scenarios using MINITAB, and response surface optimizer has been developed using response surface methodology. In this part, optimized the steam to carbon ratio, carbon dioxde to natural gas ratio, using surface response methodology using MINITAB, and the optimized conditions are translated into operator language and implemented in the systematic procedure. The implementation of the optimized conditions resulted in savings of energy 0.22 MMBTU/Ton of product and production increase of 2000 TPA.

Furthermore, a detailed dynamic model, as in the DCS of the plant, has been developed ASPEN DYNAMICS and analyzed the effects of the feed gas disturbances on the process. The analysis showed that H2/CO ratio in synthesis gas, reformer energy efficiency, methane slippage, and oxygen content in the stack gases would have serious repercussions and dynamic profiles have been developed for various feed gas compositions changes. In addition, it has been understood that a proactive advanced feed forward controls are necessary for reformer temperature, methane slippage, H2/CO ratio along with the existing feedback controls. Feed forward based control loops have been added on top of the existing controls based on the carbon number and calorific value of the feed gas. These upgradations enable the smooth operation of the plant during drastic feed gas composition changes. The added advanced process control would eliminate the situation of process upset and un-planned shutdowns, that result in improved equipment integrity and reliability.

550586 Modelling of Gas Separations By Pressure Swing Adsorption Using a Novel Steady-State Methodology

A novel pressure swing adsorption (PSA) steady-state process modeling methodology was proposed previously which greatly simplifies simulation of PSA units by using a concept of adsorption efficiency, referred to as bed-state efficiency, similar to tray efficiency in distillation. In theory, the proposed model provides a simple method for the practicing engineer to use for evaluation of PSA processes. The nature of the bed-state efficiency, however, must be elucidated before the model can see widespread acceptance. Using a combination of experimental data and dynamic simulation, an air separation unit (ASU) was investigated to gain intuition into the behavior of the bed-state efficiencies. Insights and guidance from the ASU case were then applied to the modeling of a nitrogen rejection unit (NRU) removing nitrogen from natural gas. The relationship between the bed-state efficiencies, operating conditions, and design parameters will be discussed as well as the potential for *a priori* prediction of the bed-state efficiency values.

550628 Synthesis of Reactive Separation Systems Via Building Blocks

Process intensification (PI) has become an important tool for novel process designs featuring drastic improvements in process performance. While miniaturization, use of alternative energy sources and discovery of multifunctional materials are examples for activities leading to intensification, integration of several operations in a single equipment have been one of the most successful PI strategies in terms of their industrial acceptance [1-2]. Specifically, reactive distillation (RD) units, which combine reaction and distillation operations in a single column, have seen a marked increase in their industrial adoption since 1980s [2]. Wider utilization of this technology requires systematic design procedures suggesting optimal intensified designs at the early design stage. Several methods have been suggested in the past for the synthesis of reactive distillation operations [3-4]. In this work, we use building block-based representation [5-8] to provide a systematic procedure for the synthesis and optimization of reactive separation processes. This novel representation can suggest non-intuitive intensified configurations with its unique outlook towards process design. The representation is based on collection of blocks positioned on a two-dimensional grid which can be used to represent several physical and chemical phenomena. Each building block has several physical attributes, i.e. temperature, pressure, phase, composition, and can stand for a unit use of a material, e.g. catalyst. While reaction operations can be represented within a single block, separation related phenomena require at least two blocks with different phases. Vapor-liquid equilibrium, for instance, requires one vapor and one liquid block sharing a common boundary through which phase transition can take place. When reaction operation is assigned into one of these neighboring blocks, a reactive vapor-liquid contact stage can be obtained. Collection of these building blocks in a grid formation yields building block superstructure which is modeled with a Mixed Integer Non-linear Programming (MINLP) model. Continuous variables are used to quantify the material and energy flow between the blocks and integer variables are utilized to assign different phenomena at different positions on the grid. This results in a generic representation and synthesis method for reactive separation processes. We will demonstrate the applicability of the proposed method with several example problems including ideal and non-ideal reactive systems and show that cost-optimal intensified designs can be realized through building-block based approach.

[1] Stankiewicz, A. I., Moulijn, J. A. (2000). Process Intensification: Transforming Chemical Engineering. Chemical Engineering Progress 1, 22–34.

[2] Tian, Y., Demirel, S. E., Hasan, M. M. F., Pistikopoulos, E. N. (2018). An Overview of Process Systems Engineering Approaches for Process Intensification: State of the Art. Chemical Engineering and Processing: Process Intensification, 133, 160-210.

[3] Ismail, S.R., Pistikopoulos, E.N. and Papalexandri, K.P. (1999). Synthesis of reactive and combined reactor/separation systems utilizing a mass/heat exchange transfer module. *Chemical engineering science*, *54*(13-14), 2721-2729.

[4] Okasinski, M.J. and Doherty, M.F. (1998). Design method for kinetically controlled, staged reactive distillation columns. *Industrial & Engineering Chemistry Research*, *37*(7), 2821-2834.

[5] Demirel, S. E., Li, J., and Hasan, M.M.F. (2017). Systematic Process Intensification using Building Blocks, *Computers and Chemical Engineering*, 105, 2-38.

[6] Li, J., Demirel, S.E. and Hasan, M.M.F. (2018). Process synthesis using block superstructure with automated flowsheet generation and optimization. *AIChE Journal*, 64(8), 3082-3100.

[7] Li J., Demirel S.E., Hasan M.M.F. (2018). Process Integration using Block Superstructure. *Industrial & Engineering Chemistry Research*, 57, 4377–4398.

[8] Li J., Demirel S.E., Hasan M.M.F. (2018). Fuel Gas Network Synthesis Using Block Superstructure. *Processes*. 2018, 6(3), 23.

550417 A Comparative Study of Residence Time Distribution and Cooling Crystallization in a Continuous Dynamic/Oscillatory Baffle Crystallizer Versus a Stirred Tank

Crystallization has become a predominant technique in particle design technology and it is widely applied in the pharmaceutical industry. Traditionally crystallization has been carried out in batch mode where operating conditions vary with time, often resulting in inconsistent products.[1] Continuous crystallization, on the other hand, has been identified to improve reproducibility, reduce operating cost and simplify scale-up.[2] A novel commercial dynamic baffle crystallizer(DBC) operated as a mixed-suspension-mixed-product-removal(MSMPR) system has been studied. It consists of a reactor tank and dynamic 'donut' shaped baffles to provide oscillatory flow resulting in more uniform mixing.[3] Oscillatory systems have been well studied for batch reactions and synthesis but few studies have been carried out to study the DBC for continuous two-phase operations such as crystallization.

In the work proposed here, liquid and solid residence time distributions were studied systematically in the DBC in comparison with a traditional continuous stirred tank crystallizer(STC). In-situ process analytical technology(PAT) tools such as UV-vis, Infrared(IR), and focused beam reflectance measurement(FBRM) were used to measure system response to step or pulse changes in order to construct RTDs. It is observed that the uniformity of both liquid and solid RTDs was significantly improved in the DBC than the STC indicating potential improvement in size uniformity during crystallization processes. A preliminary continuous cooling crystallization of paracetamol was carried out for two different oscillating conditions in the DBC as well as the STC at the corresponding rotational speeds yielding (approximately) equal power density. Online PAT tools were used to monitor the processes in addition to sampling for offline characterization at every residence time (45min). Nucleation event took place earlier in the DBC than the STC possibly due to the improved mixing mechanism. At early stages of operation, DBC produced crystalline particles of comparable or improved size uniformity. At later stages of operation (>7 residence times), particle sedimentation or severe particle breakage occurred in the STC causing process deviation from steady state while such events did not take place in the DBC and steady state was maintained. In the RTD studies as well as the preliminary crystallization study, the DBC showed improved system uniformity and consistency demonstrating potentials for more in-depth crystallization studies in the DBC.

1. Nagy ZK, Aamir E. Systematic design of supersaturation controlled crystallization processes for shaping the crystal size distribution using an analytical estimator. *Chem Eng Sci*. 2012;84:656-670. doi:10.1016/j.ces.2012.08.048.
2. Alvarez AJ, Singh A, Myerson AS. Crystallization of Cyclosporine in a Multistage Continuous MSMPR Crystallizer. *Cryst Growth Des*. 2011;11(10):4392-4400. doi:10.1021/cg200546g.
3. Hewgill MR, Mackley MR, Pandit AB, Pannu SS. Enhancement of gas-liquid mass transfer using oscillatory flow in a baffled tube. *Chem Eng Sci*. 1993;48(4):799-809. doi:10.1016/0009-2509(93)80145-G.

546606 Design of Vapor Recompressed Dividing-Wall Columns Processing a Light Component Dominant Mixture

When a dividing-wall column (DWC) processes a mixture with a dominant composition of the light component, there exists not only a relatively small temperature elevation from the overhead to the stripping section of its prefractionator but also a large amount of latent heat available in the overhead vapor. These provide opportunities for applying heat pump to the DWC and lead potentially to great improvement in system performance. In this paper, a DWC processing an N-pentane/N-hexane/N-heptane ternary mixture is chosen to examine the characteristics and effect of the applications of heat pumps. We use Stage-Enthalpy Deficit (S-H) profile and Temperature-Enthalpy Deficit (T-H) profile to guide the combinations of heat pumps with the DWC while taking the minimization of total annual cost (TAC) as an objective function. The obtained result shows that multiple heat pumps might be employed to use the compressed overhead vapor to heat the feed, the stripping section of the prefractionator, and the section above the intermediate product of the main column. The first heat pump is employed to preheat the feed, thereby allowing the adoption of feed splitting to strengthen the contact between vapor and liquid phase in the prefractionator. The second heat pump is installed to heat the stripping section of the prefractionator and the third heat pump is installed to heat the separating section above the intermediate product of the main column. The TAC decreases by 25.09% as compared with the DWC. The total utility consumption (TUC) decreases by 62.55% as compared with the DWC. These striking results highlight the great potential of applying heat pump to facilitate the DWC. Current work is underway to simplify the search process in process synthesis and design.

551422 Theoretical Modeling of Mixing Quality and Axial Hold-up Distribution for Liquid-Liquid System in Agitated Flow Mixers

New theoretical based models to describe the mixing quality and the local dispersed phase hold-up distributions in the axial direction of a continuous flow mixer have been developed and applied to predict the experimental results. The models relate the mixing index and axial hold-up distributions to a Peclet Number, the average dispersed phase hold-up, the input flow rates and the physical properties of the dispersion. The developed models offer a new understanding of the nature of the mixing process in continuous mixers with two liquid-liquid immiscible phases in flow. Experimental work was performed to measure the mixing index and to measure axial dispersed phase hold-up distribution using Laser Transmission Technique (LTT) in a mixer setup. The experimental results were used in validating the models. The experimental and the predicted data agreed well for a wide range of impeller speeds/flow ratios. Theoretical formulae to describe the local hold-up distributions in the axial direction for only one of the two phases in batch and in flow operation processes have also been developed.

Paper presented at AIChE Spring Meeting,
Final Paper #550370
March 31 – April 4 2019, New Orleans, USA Ver.1, 28/3/19

DEPLOYMENT OF DIGITAL MODELLING FOR CRACKING FURNACE ONLINE YIELD OPTIMIZATION

Yasser Al-Goufily[1], Abdullah Alqahtani[2], Abduljelil Iliyas[3], and Steve Hall[4]

(1) SABIC Technology Centre, Al-Jubail, Saudi Arabia,

(2) SABIC Technology Centre, Saudi Basic Industries Corporation, Riyadh, Saudi Arabia,

(3) SABIC Technology Centre, Saudi Basic Industries Corporation, Houston, USA

(4) Process Systems Enterprise Ltd., London, United Kingdom.

Contact: Yasser Al-Goufily, Tel: + 966 13 359 9166, Email: 29574@sabic.com

1. Introduction

Olefins plants are at the heart of the petrochemicals value chain. They convert gas and liquid feedstocks such as ethane and naphtha to ethylene, propylene and butylene. These products are then further processed into a wide range of valuable performance chemicals such as polymers, resins, esters, glycols and solvents. This conversion is performed at high temperature in thermal cracking furnaces. During the cracking process, coke is deposited in the furnace coils and transfer line exchanger.

An olefins plant is the single most energy-consuming process in the chemical industry. Modern plants have capacities up to 1.5 million tonnes/year ethylene and at this scale, small improvements in efficiency can yield significant annual savings. Olefin plant operation is thus a prime candidate for optimisation aimed at maximising economic performance.

There are several challenges that operators face regarding cracking furnaces (a) setting the desired yield for each furnace based on overall optimality, (b) understanding the extent of coking in the furnace which has a direct effect on residence time and tube metal temperatures, (c) accurately predicting ethylene yields - a key measure of plant performance and an important variable for severity control.

Another over-riding challenge is to implement improvements as part of a digital strategy, where the information contained within plant data is fully exploited by a model and the calculated results can be shared with other applications.

This paper describes how SABIC has implemented such a digitally-integrated system. The model calculates the key cracking furnace variables including conversion, product yields and degree of coking. The coking calculation uses historical plant data and a state-estimation technique to provide significantly better estimates for plant performance indicators than previous approaches. The data and calculated results are also available for other applications as needed.

2. Challenges to Operations

There are several challenges that operators face regarding cracking furnaces.

2.1 Setting the desired yield

A major challenge for operators is to determine what is the target yield in each furnace, given the fact that feed flowrates may be varying, as may the feed compositions. Also, as the proportion of ethane, propane and naphtha in the feeds varies, the number of furnaces dedicated to each feedstock type may also have to change to maintain overall economic optimality. Finally, the price of each feedstock together with prices of products and utilities may also vary. Together, all these will affect the optimal conversion to be set for each furnace.

2.2 Understanding extent and effects of coking

Coking starts from the first day of coil operation and continues until the effects are serious enough to warrant taking the unit out of service for de-coking. The degree of coking affects virtually all aspects of operation and an accurate, up-to-date prediction of coking is a critical pre-requisite for optimisation and yield accounting. The rate of coking can be exacerbated by poor operation. The effects of coke build-up increase over time and include reduced heat transfer leading to higher metal skin temperatures, increasing pressure drop and, in general, lower efficiency of operation with more energy required per unit of production. This leads to loss of yield and shortened run length. More serious is during upset operation, for example power failure, where there is significantly higher potential to damage the cracking tubes with resulting lost production and replacement costs that can run to tens of millions of dollars. Such damage can also impact safety.

2.3 Determining accurate yield prediction

The lack of accurate and timely yield information can have a significant impact on the ability to control furnace operation, leading to non-optimal operation. Given the immense cash flows in olefins plant operations, any non-optimal operation can represent a substantial missed revenue or cost-saving opportunity. An accurate measure of conversion is critical for effective furnace severity control. However, measuring the conversion and product yields from furnaces is a constant challenge because physical instruments which use extractive sampling techniques such as process gas chromatographs for composition measurement often fail or are subject to fouling. This can result from tars and contaminants created in the furnaces which can contaminate the sample lines and analyser. These affect sampling rates and sample conditioning and as a consequence effluent analysers are frequently out of order or poorly calibrated. In most cases it is not easy to maintain analysers during normal operation. In addition, each furnace will likely have its own analyser and there may, in the case of ethane/propane mixed feed furnaces, be feed analysers too. All these analysers provide many potential sources of error.

3. Simulation-based modelling approaches

There are several existing simulation-based approaches that attempt to *predict* rather than *measure* yield in real-time. Typically, yield models might use feed flow and composition, coil inlet pressures and temperatures. Often there is a limited attempt to take into account the degree of coking as it occurs throughout the production run and because coking affects conversion, the results regularly have to be ‘interpreted’ by experienced operators. In addition, uncertainty in any of the measured variables from the plant is not taken into account, mainly because there is no way using conventional approaches to incorporate such uncertainty into any predictive calculations. Hence there is a risk of inaccurate values being calculated for critical parameters such as yield, pressure drop and maximum tube metal temperature,

Sometimes yield models are used in conjunction with coking models, with coking over the run length simulated by a semi-empirical coking model using presumptive coking kinetic parameters fixed beforehand. The available models do not generally predict coking accurately, especially given that coke build up varies radially, axially and over time. Coke predictions tend to depart from reality over the normal course of a run, particularly if there is deviation from standard operation. Consequently such models lack the ability to predict yield and tube metal temperature to a sufficient degree of accuracy.

In some cases, soft sensors based on statistical models[1] are used. Again, these reflect 'normal' operation but have no predictive capability, so can deviate from reality if there is some type of upset in the operation.

4. Advanced process modelling approach for accurate yield prediction and coking monitoring

The digital model-based approach described in this paper is based on a novel Advanced Process Modelling (APM) methodology[2] that couples detailed predictive furnace models with overall olefins plant models and regularly-updated plant data.

4.1 Determining desired furnace yields

The following figure shows the scope of the overall olefins plant model. It covers not only cracking furnaces, but also quench, compression and downstream separation. All these have an impact on operating costs.

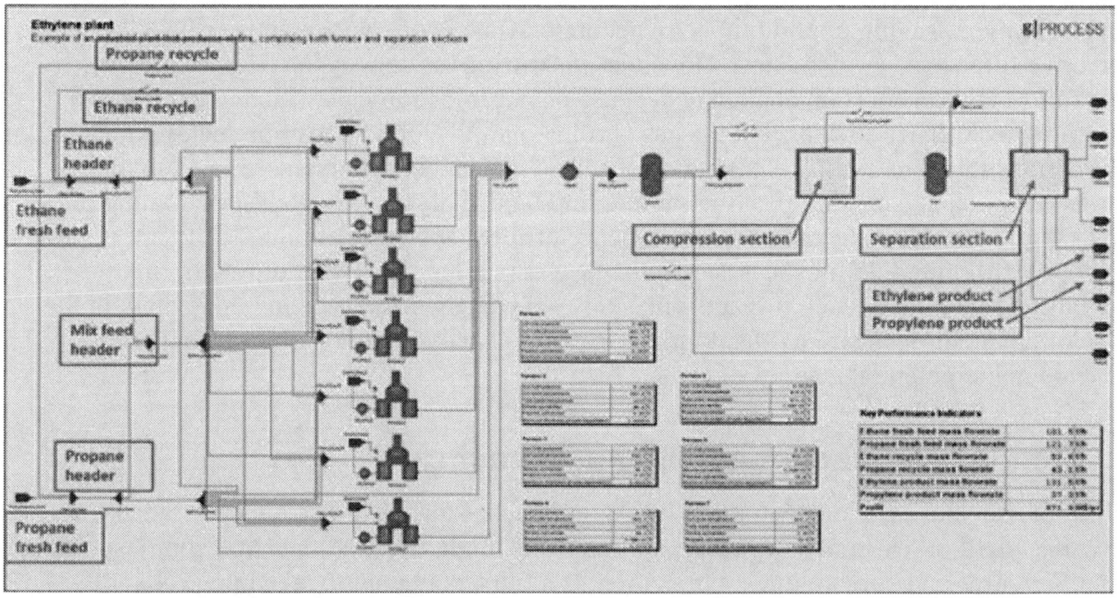

Figure 1 – Overall olefins plant model

---

[1] Soft sensors for processing plants, by G. D. Gonzalez, published in the Proceedings of the Second International Conference on Intelligent Processing and Manufacturing of Materials, July 10-15, 1999, Honolulu, Hawaii

[2] First-principles modelling for optimization of process operations, Costas Pantelides, Foundations of Computer Aided Process Operations / Chemical Process Control, Tucson, Arizona, January 8-12, 2017

The predictive cracking furnace model includes first-principles representation of coil geometry, material and heat balances, cracking kinetics and coking kinetics. The plant data used by the model are measured variables such as feed rate, feed composition, coil inlet temperature and coil outlet temperature, together with any additional available measurements that can be used for real-time calibration of the model, such as pressure drop over coils or tube metal temperatures. Uncertainty is assigned to these additional measured variables and these are considered in the overall reconciliation of the plant data.

In addition to feed flows and compositions, feed, product and energy pricing is considered in the overall economic model. The whole plant model runs and determines the desired yield required in each furnace. The model can be run regularly or whenever feedstocks or pricing change.

4.2 Determining the degree of coking

The furnace system models are solved using dynamic state estimation techniques based on an Extended Kalman Filter approach to provide up-to-date estimates of process state variables, selected process parameters and other key performance indicators (KPIs). The use of Kalman filtering as feedback in process control is not new. However, the novelty here comes from combining the use of Kalman filters with rigorous process simulation models.

The approach exploits data redundancy features within the data system and the applied confidence limits to measured variables to reduce overall model uncertainty, i.e. the approach makes maximum use of all the data available to it. This powerful technique allows the model to be used, if necessary, beyond its initial operating window. Kinetic parameters of the coking model are adjusted periodically to match observed reality as new evidence arrives in the form of actual plant measurements such as pressure drop over the cracking coils. Effectively this results in a *self-calibrating coking model* that can accurately predict the current state of coking along the length of the coil and take this into account in the yield prediction calculation.

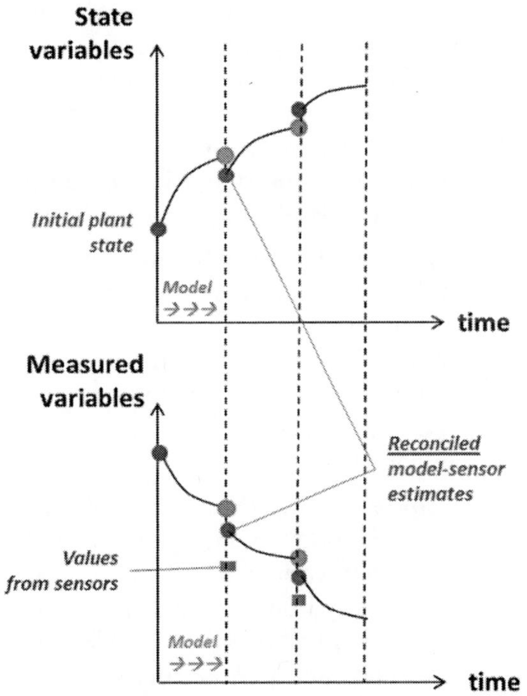

Figure 2 – State estimator operation

The operation of a state estimator in conjunction with a process model is illustrated in Figure 2. At the beginning of the first interval, the plant model has what it believes are correct values for the plant state (e.g. pressures, state of coking) and various measurements. It then calculates new values for the next time interval (orange dots) based on this understanding. There is an element of error in the new prediction, because no model is ever perfect.

In the meantime, new measurements (grey rectangles) arrive from the plant sensors. These will typically differ from the values calculated by the model; they also carry a degree of uncertainty, reflecting, for example, instrument error.

The state estimator then uses mathematical and statistical techniques to find the best 'reconciled' values for all the variables that it is allowed to adjust (blue dots). This effectively simultaneously takes into account uncertainty in plant measurements as well as possible errors in the model representation (for example, resulting from an inaccurate coking model), to provide a reconciled plant model. This process continues for each subsequent time-interval.

At the end of each interval, the model contains the best up-to-date reconciled information that can be achieved for the process, subject to the accuracy of the model representation and the uncertainty in the measurements.

4.3 Determining furnace yields

In order to be deployed effectively by the state estimator, the underlying furnace model requires a certain minimum level of fidelity to calculate accurate yields. In particular it is important that it has a high degree of *predictive* capability. The model needs to include, at a minimum:

- **Furnace geometry.** The coil geometry needs to be represented accurately, along with its material properties such as thermal conductivity.
- **Heat input to cracking coils.** It was found in offline studies that using a representative relative heat release profile according to the furnace design together with a severity specification provides sufficient fidelity on modelling the firebox side of the furnace when the focus is on prediction of cracking yields and coking rates along the tubes. Imposing a relative heat flux profile can therefore be used as a much simpler and faster approach that is very effective for representing the actual heat input, as opposed to using more rigorous methods such as zonal methods or computational fluid dynamic-like approaches.
- **Detailed cracking kinetics**. Accurate kinetics for the cracking reactions are critical to reliable modelling of the furnace and indeed the whole ethylene plant performance. There are many ways of supplying kinetics: kinetic calculations supplied with the modelling package; commercially-available kinetics packages; or via users' own kinetics, perhaps fitted to plant or laboratory data. In the modelling tool used, supplied kinetics based on literature sources can be adjusted to match observed plant or pilot plant data.
- **Coking model**. Again, the coking model is important as coking starts from day one of a production run and affects virtually all aspects of operation and economics. Given the many variables in furnace operation, it is virtually impossible to predict kinetics accurately from a static model; this is a key reason for the approach described here. Well-established kinetic models from literature were used, with parameters such as rate constants periodically updated based on observed plant performance, using dynamic state estimation techniques.

A furnace model fulfilling these requirements was implemented in SABIC under the name 'gCRACKER' (also referred to by PSE as gPROMS Olefins Cracking Monitor). Figure 3 shows some of the furnace configuration screens related to coil selection and definition of the coking kinetic model.

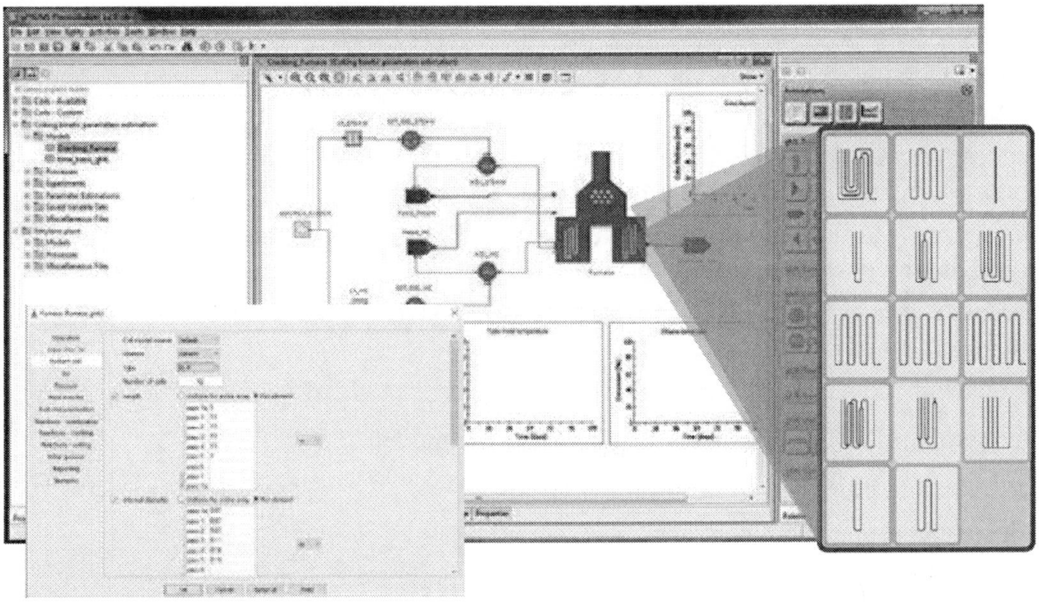

Figure 3 – Ethylene furnace setup showing coking kinetics definition dialogs

5. Experience in SABIC crackers

The technology has been applied to a large ethylene plant in Saudi Arabia processing mainly ethane and propane feedstocks.

5.1 Workflow

In implementing the gCRACKER model in SABIC's plants, a number of key elements are included in the workflow, as represented in the following figure.

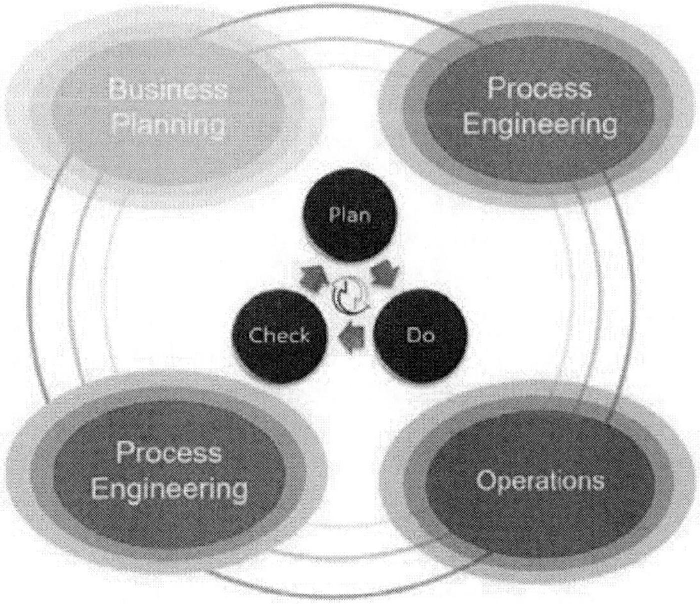

Figure 4 – Overall workflow to deploy a digital olefins plant model in SABIC

The various workflow elements are as follows:

- Business planning – to provide a weekly run plan considering the required production volumes and costs
- Process engineering - to utilize the gCRACKER optimization model and determine the operating parameters which maximize profitability
- Operations – to run the furnaces with gCRACKER at the optimized conversion and highlight any operating limitations
- Process engineering – to estimate the benefits considering the contribution margins of the generated products

These various workflow elements ensure that the best use is made of the model.

The following optimization parameters were used in the model:

- Ethylene and propylene contribution margins
- Ethane and propane conversions
- Ethane and propane recycle flows

5.2 Results

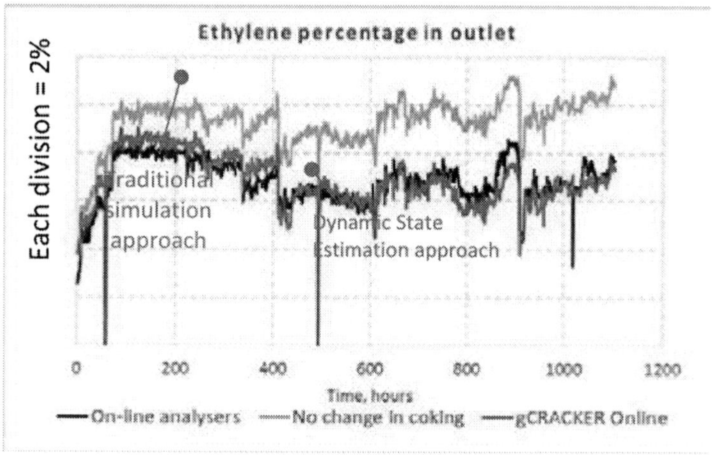

Figure 5 – Ethylene prediction (red and grey line) compared with plant data (black line)

The plot in Figure 5 shows results from both gCRACKER (CrackingMonitor) and a simulation case, compared to plant data. The data were taken from a period of operation where reasonably reliable ethylene yield measurements were available and could therefore be used to provide verification of model predictions.

During this performance run, great care was taken to ensure the relevant analysers were working well, in a calibrated state with a working sampling system. Two sets of results are compared against the on-line analyser data (black line) over a 50-day run.

The grey line shows the ethylene yield when the coking is simulated assuming fixed coking parameters and no change to coking rate. The conditions in the furnace are simulated as the run progresses, taking into account actual feed flowrate, controls, etc., to determine coking and ethylene yield. This approach, which is used by current commercial software tools, works well in the early stage of the run. However, the inability of the coking model to accurately reflect all events affecting coke deposition over the course of the run eventually causes the calculated coking state and hence ethylene yield prediction, to deviate significantly from reality.

By contrast, the red line shows the prediction using the dynamic state estimation techniques. This is more accurate than simply simulating the operation, as the plant data are used to adjust the prediction of coking progressively as the run unfolds – i.e. the coking model is effectively *self-tuning*, meaning that the degree of coking reported is always accurate and up-to-date.

Note that the ethylene yield measurement recorded by the on-line process analysers was *not* included in the state estimation approach so the ethylene concentration is a purely *predicted* value. It can be seen that the predictions reflect the actual measured values very closely.

5.3 Benefit realization

Having determined the optimal conversion required in each furnace through the overall olefins plant optimizer, the closed-loop on-line Gcracker (CrackingMonitor) tool was employed. The optimization parameters were set and the conversion set-points relayed to the APC controllers.

The realized benefit of online optimization was an increase in olefins production by **2.8** tons / hour which translates to a yield enhancement of **2%**.

Ethylene Benefit (tph)	+ 0.3
Propylene Benefit (tph)	+ 2.5
Olefins Yield Increase %	**+ 2%**

Table 1. Benefits of implementing the gCRACKER (CrackingMonitor) tool

6. Conclusions

With the whole olefins plant model and optimizer, linked to furnace model and state estimator, all connected to the plant's data acquisition system, plant data is processed in real time. This digital approach provides:

- highly accurate predictions of conversion and product yields, which can be used for furnace severity control
- accurate estimates of the current degree of coke deposition in each section of the furnace coil, for use in optimising de-coking intervals
- accurate soft-sensed measurements of furnace KPIs or any other variables of interest in the plant that can be used for reporting or for improved control
- an up-to date model of each furnace that, when included with a high-fidelity model of the surrounding process incorporating updated plant parameters, can be used for whole-plant optimisation (online or offline), look-ahead simulation and many other activities.

In terms of benefits, this results in:

- significantly improved confidence in the reliability of furnace effluent composition values

- the ability to accurately control the furnace based on conversion as opposed to, for example, Coil Outlet Temperature
- improved furnace control, leading in turn to higher throughput and profitability
- better control and management of coking, leading to higher efficiency and longer run lengths
- early warning of excessive coking, helping to avoid split tubes and other unplanned outages which can be very costly
- better, up-to-date process models on which to base optimisation, leading to better, more profitable operation
- accurate soft-sensed values for all variables within the scope of the process model, leading to greater understanding and monitoring of operation
- improved consistency between off-line and on-line models which aid in accurate predictions for production planning and control.

A key further application of the approach outlined here is in providing an up-to-date plant model – including accurate coking prediction and adjusted reaction kinetics – for a real-time plant optimiser. This is a crucial requirement for extracting the maximum value from online optimisation calculations.

The realized benefit of online optimization was an increase in olefins production by **2.8** tons / hour which translates to a yield enhancement of **2%**.

551016 Optimization of Mixed Smr Plant Using Advanced Modelling & Six Sigma Tools to Enable Smart Manufacturing: Case Study

Synthesis gas is the key feedstock for many industrial applications especially for ammonia, methanol, Oxo-alcohols etc. Most of the synthesis gas produced in the world is from natural gas using steam methane reforming (SMR). Mixed SMRs have carbon di-oxide added to the feed gas to produce synthesis gas with lower H2/CO ratio than that of conventional synthesis gas applications. As the gas reserves are depleting in the world, existing reforming plants are open to use natural gas from different sources. Different natural gas sources offer different compositions in the gas, which significantly affect the performance of the mixed steam methane reformers (SMR) in terms of reforming efficiency, energy consumption, H2/CO ratio in synthesis gas etc. Effect of feed gas disturbances in mixed SMRs is significant as these plants are operated at higher reforming temperature and lower steam to carbon ratios. In order to operate the reforming plant efficiently, optimally and smartly for all feed gas disturbances, optimization of the plant using advanced modelling & six sigma tools, precise pro-active control, continuous operations monitoring are necessary.

In this paper, an implemented optimization of mixed steam reformer in the oxo-alcohol production process case study is presented. An Aspen Plus model based study has been performed to optimize the mixed SMR plant operations for the existing feed gas composition. Then, coupled the model-based study with six sigma tools to develop the best operating conditions and zone for various scenarios. A number of contour diagrams have been developed for these scenarios using MINITAB, and response surface optimizer has been developed using response surface methodology. In this part, optimized the steam to carbon ratio, carbon dioxde to natural gas ratio, using surface response methodology using MINITAB, and the optimized conditions are translated into operator language and implemented in the systematic procedure. The implementation of the optimized conditions resulted in savings of energy 0.22 MMBTU/Ton of product and production increase of 2000 TPA.

Furthermore, a detailed dynamic model, as in the DCS of the plant, has been developed ASPEN DYNAMICS and analyzed the effects of the feed gas disturbances on the process. The analysis showed that H2/CO ratio in synthesis gas, reformer energy efficiency, methane slippage, and oxygen content in the stack gases would have serious repercussions and dynamic profiles have been developed for various feed gas compositions changes. In addition, it has been understood that a proactive advanced feed forward controls are necessary for reformer temperature, methane slippage, H2/CO ratio along with the existing feedback controls. Feed forward based control loops have been added on top of the existing controls based on the carbon number and calorific value of the feed gas. These upgradations enable the smooth operation of the plant during drastic feed gas composition changes. The added advanced process control would eliminate the situation of process upset and un-planned shutdowns, that result in improved equipment integrity and reliability.

553432 Molecular Analysis Reveals the Myth of Value Chain and Maximize Potential

Process synthesis is about finding the appropriate processing routes with selection of catalyst, reaction and separation technologies followed by flowsheeting (connection of selected technologies) and equipment design at the least capital and operating cost. The common feature of current process synthesis methods is that they are mainly based on macroscopic parameters but do not exploit microscopic effects at molecular level. As a result, the fundamental aspects cannot be fully exploited to discover novel ideas. However, the essence of process synthesis is about maximizing the production of desirable chemical species via optimizing the routing of key species to their most appropriate destinations. To achieve this goal, a molecular analysis methodology for process synthesis is proposed. Several industrial case studies are provided to prove the concept and demonstrate the value.

550027 Thermodynamic Modeling of Lithium Salts with the Electrolyte NRTL Model Using Hydration Chemistry

The lithium bromide-water solution is a popular working fluid due to its environmentally friendly, nonvolatile, and non-toxic properties; however, this solution has issues with corrosion and crystallization. One proposed solution is to modify the lithium bromide-water working solution by adding anticrystallization and anticorrosion agents such as lithium chloride, lithium iodide, and lithium nitrate. Thermodynamic models are essential to facilitate the advancement of process design, optimization, and operation range of absorption heat pumps and refrigeration units that utilize lithium salt systems. To obtain a rigorous thermodynamic model of lithium salt solutions, appropriate solution chemistry must be established. The current work proposes that the lithium ion is present with permanently bound water molecules that should be considered as part of the ion. The hydration chemistry provides an increased accuracy to the symmetric electrolyte NRTL (eNRTL) activity coefficient model. This solubility model correlates the nonideality of the solution with the composition dependency by utilizing two binary interaction parameters for each of the electrolyte-electrolyte and molecule-electrolyte pairs in the system. Each parameter is determined by applying up to three temperature dependent coefficients and then regressed from experimental data. This allows the parameters to establish the accuracy of the model for concentrations up to saturation and temperatures between 273.15-473.15 K.

550607 Multiscale Process Simulation By Combining Subgrid Modeling and Computational Fluid Dynamics

Many realworld processes are multi-scale processes, which means that they have important features at different scales in time and/or space. Typically not all features can be resolved, so that for simulations of such processes certain assumptions or simplifications have to be made and so--called subgrid models are used to capture the effects of one scale onto the other. The benefit of combining different scales was previously shown e.g. for a wet granulation process [1].

In this contribution we show a new automated coupling method between the simulation tools STARCCM+ by Siemens PLM and gPROMS by PSE, where STARCCM+ solves for all flow properties and gPROMS evaluates the fine scales based on the solution provided by STARCCM+. The coupling between both codes is done in an automatic way. This includes the domain decomposition, the calculation of and the transfer of all relevant variables.

The details of the coupling are discussed and the benefit of this approach is demonstrated based on used cases like a liquid-liquid emulsion for PVC production in a full scale slim vessel and a spray drying example. The numerical results are compared against experimental results [2] and the effect of different ways of decomposing the domain is discussed.

References:

1. Barosso, D. et al.: A multiscale, mechanistic model of a wet granulation process using a novel bidirectionalPBM–DEM coupling algorithm, Eng. Sci. 2015, 123, 500–513
2. Maass, S. et al.: Flow Field Analysis of Stirred Liquid-Liquid Systems in Slim Reactors, Chem. Eng. 2011, 34, No. 8, 1215–1227

550382 Dynamics and Control of Reactive Distillation Column with External Recycle

Reactive distillation columns with external recycle between the top and bottom (RDCs-TBER) have a better steady-state performance as compared with conventional reactive distillation columns (CRDCs) for the separations of reacting mixtures featuring the most unfavorable ranking of relative volatilities (i.e., the reactants are the lightest and heaviest components with the generated products in between). In this article, their dynamics and control are studied in great detail, with special attention given to the influences of external recycle on process dynamics and controllability. Three kinds of control systems are developed. The top and bottom levels of all control systems are controlled by reflux flow rate and reboiler heat duty, respectively. The only distinguish of these three control systems is that one is with constant external recycle, another is with constant reflux ratio, and the last is controlling the product purity with external recycle flow rate.

Two reactive distillation systems, executing a hypothetical reversible reaction, $A + B \leftrightarrow C + D$ ($\alpha_A > \alpha_C > \alpha_D > \alpha_B$), and lactic acid esterification, are employed to inspect the dynamics and controllability of the RDC-TBER. The results show that the supplementary manipulated variable (i.e., the flow rate of the external recycle) can be used to improve the process operation. The RDC-TBER, on the other hand, shows greatly improved tracking performance with even enhanced disturbance rejection capabilities.

549570 Separation of Dmc/Methanol Azeotrope By Heat Integrated Distillation Equipped with Pervap Membrane

E3Tec has developed Heat Integrated Reactive Distillation (HIRD) process equipped with side reactors and PerVaporization (PerVap) membrane for conversion of CO_2 to dimethyl carbonate (DMC). Separation of DMC/methanol azeotrope for achieving high-concentration (99.9%+) DMC is energy intensive. E3Tec has developed an integrated distillation for separation of DMC/methanol azeotrope by recovery of the overheat latent heat, internal thermal integration and PerVap integrated with the distillation column. The resulting DMC purity of 99.9%+ was achieved as bottom product and DMC in the overhead methanol-rich stream is reduced. Methanol is recycled in the CO_2 conversion process as feed to side reactors; therefore, minimizing DMC content in the recycled methanol stream is preferred for a reversible, equilibrium- controlled reaction. The analysis was based on the ASPEN Plus® simulation of the integrated distillation subsystem. The project was supported by DOE SBIR program.

549973 High-Throughput and High-Efficiency Liquid-Liquid Extraction Column WINTRAY -Review of the Commercial Experiences and Recent Development-

The WINTRAY®, the liquid-liquid extractor, was commercialized to obtain high-throughput high-efficiency and scum-resistant in the liquid-liquid extraction. It has been used in 19 commercial plants worldwide.

For new construction, the size of the column becomes compact by using WINTRAY® because of its high throughput and high efficiency. As compared to extractors coming with driving device, the simple structure which consist of trays and weirs with openings, leads to low capital cost and maintenance cost.

For intensification of existing extraction columns, throughput can be increased by replacing the internals with WINTRAY®. And also, the energy consumption in the solvent recycle section can be saved by reducing the solvent circulation rate of the extraction column.

We herein report the characteristics and commercial examples of process intensification using WINTRAY® . In addition, the result of the further improvement to increase the throughput are introduced.

550485 Material Selection for Packed Columns in the Chlorine Drying Stage of the Chlor-Alkali Process

The Chlor-alkali industry uses electrolytic processes to produce chlorine and caustic soda from aqueous sodium chloride solutions. Chlorine drying is one of the stages of chlorine processing which removes moisture from chlorine by using sulfuric acid. Two or more columns are employed in series for chlorine drying by counter-current contact of gas flow with acid flow in a packed column. For the latter, material selection is pivotal to the process design because even small traces of water in chlorine can be highly corrosive. Water content in chlorine and sulfuric acid drive the material selection, which in turn is dependent on each material's chemical resistance, temperature rating, mechanical strength and cost. Design and economic considerations that lead to selecting materials for the column shell, internals and high capacity random packing are discussed in this paper. Quantitative comparisons of the different packing materials in terms of their hydraulic properties will be shown. This will be supported using select commercial application examples.

549941 Using Ethanol in Continuous Biodiesel Production with Trace Catalyst and CO_2 Co-Solvent

ABSTRACT

The continuous biodiesel production process under sub and super-critical conditions using a trace amount of potassium hydroxide (KOH) as a catalyst have been studied. In addition, CO_2 was added as a co-solvent to reduce the reaction time and increase biodiesel yield. The proposed procedure enables simultaneous transesterification and esterification of triglycerides and free fatty acid (FFA), respectively. The shorter reaction time and milder reaction conditions may reduce energy consumption due to simplification of the separation and purification steps. The process variables, including reaction temperature, ethanol to oil molar ratio, catalyst amount, and process pressure were systematically optimized. The highest biodiesel yield (98.12%) was obtained after 25 minutes reaction time using only 0.11% wt. of KOH and 20:1 ethanol to oil ratio. The process optimum temperature and pressure were 240 °C and 120 bars, respectively. The proposed kinetic model suggested a first order reaction with an activation energy of 19.5 $kJ.mol^{-1}$ and a reaction rate constant of $0.0398/min^{-1}$. The thermodynamic parameter values including enthalpy, entropy and Gibbs free energy for this process were calculated as 15.05 $kJ.mol^{-1}$, -0.27 $kJ.mol^{-1}$ and 153.6 $kJ.mol^{-1}$ at 240 °C, respectively.

AUTHOR INDEX

AUTHOR INDEX

AIChE
120 Wall Street, FL 23
New York, NY 10005-4020

ISBN 978-1-5108-8696-4

CURRAN ASSOCIATES INC.
proceedings.com